PRENTICE-HALL INTERNATIONAL SERIES IN MANAGEMENT

Athos and Coffey	Behavior in Organizations: A Multidimensional View
Baumol	Economic Theory and Operations Analysis
Boot	Mathematical Reasoning in Economics and Management Science
Brown	Smoothing, Forecasting and Prediction of Discrete Time Series
Chambers	Accounting, Evaluation and Economic Behavior
Churchman	Prediction and Optimal Decision: Philosophical Issues of a Science of Values
Clarkson	The Theory of Consumer Demand: A Critical Appraisal
Cohen and Cyert	Theory of the Firm: Resource Allocation in a Market Economy
Cyert and March	A Behavioral Theory of the Firm
Fabrycky and Torgersen	Operations Economy: Industrial Applications of Operations Research
Greenlaw, Herron, and Rawdon	Business Simulation in Industrial and University Education
Hadley and Whitin	Analysis of Inventory Systems
Holt, Muth, Modigliani, and Simon	Planning Production, Inventories, and Work Force
Hymans	Probability Theory with Applications to Econometrics and Decision-Making
Ijiri	The Foundations of Accounting Measurement: A Mathematical, Economic, and Behavioral Inquiry
Kaufmann	Methods and Models of Operations Research
Lesourne	Economic Analysis and Industrial Management
Mantel	Cases in Managerial Decisions
Massé	Optimal Investment Decisions: Rules for Action and Criteria for Choice
McGuire	Theories of Business Behavior
Miller and Starr	Executive Decisions and Operations Research (2nd ed.)
Montgomery and Urban	Management Science in Marketing
Morris	Management Science: A Bayesian Introduction
Muth and Thompson	Industrial Scheduling
Nelson (ed.)	Marginal Cost Pricing in Practice
Nicosia	Consumer Decision Processes: Marketing and Advertising Decisions
Peters and Summers	Statistical Analysis for Business Decisions
Pfiffner and Sherwood	Administrative Organization
Simonnard	Linear Programming
Singer	Antitrust Economics: Selected Legal Cases and Economic Models

Vernon	Manager in the International Economy
Wagner	Principles of Operations Research with Applications to Managerial Decisions
Zangwill	Nonlinear Programming: A Unified Approach
Zenoff and Zwick	International Financial Management

PRENTICE-HALL, INC.
PRENTICE-HALL INTERNATIONAL, INC., UNITED KINGDOM AND EIRE
PRENTICE-HALL OF CANADA, LTD., CANADA
J. H. DEBUSSY, LTD., HOLLAND AND FLEMISH-SPEAKING BELGIUM
DUNOD PRESS, FRANCE
MARUZEN COMPANY, LTD., FAR EAST
HERRERO HERMANOS, SUCS., SPAIN AND LATIN AMERICA
R. OLDERBOURG VERLAG, GERMANY
ULRICO HOEPLI EDITORS, ITALY

Nonlinear

Programming

A Unified Approach

by WILLARD I. ZANGWILL

Associate Professor, School of Business Administration,
University of California, Berkeley

Nonlinear Programming

Programming

A Unified Approach

PRENTICE-HALL, INC., ENGLEWOOD CLIFFS, N. J.

The example in Sec. 1.4 and problems 3.4 and 3.5 are adapted from Duffin, Petersen, and Zener, *Geometric Programming*. New York: John Wiley & Sons, Inc., 1967.

© 1969 PRENTICE-HALL, INC., ENGLEWOOD CLIFFS, N.J.

Current printing (last digit)
10 9 8 7 6 5 4

13–623579–4

Library of Congress Catalog Card Number: 69–10606

Printed in the United States of America

PRENTICE-HALL INTERNATIONAL, INC., *London*
PRENTICE-HALL OF AUSTRALIA PTY. LTD., *Sydney*
PRENTICE-HALL OF CANADA, LTD., *Toronto*
PRENTICE-HALL OF INDIA PRIVATE LTD., *New Delhi*
PRENTICE-HALL OF JAPAN, INC., *Tokyo*

to Judy, Michael, and Monica

Preface

Nonlinear programming is a pivotal area, for such problems abound in the world. This book presents a new and unified approach that not only simplifies the learning task but also develops many of the latest research results. Indeed, most of the book is based upon the author's and others' current research and has never appeared elsewhere.

Although nonlinear programming resembles linear programming, the functions need not be linear, even though they still could be. The added generality of nonlinear functions permits extremely accurate modeling of real-world problems. This book discusses both the formulation and solution of these problems.

A key question is whether a particular algorithm can actually calculate or, more precisely, converge to the solution of a given problem. A sizable portion of this book analyzes this question and takes the approach suggested by the author's research into the theory of algorithmic convergence. Thus far, this theory has simplified many previously known proofs, has proved convergence of new algorithms that the developers were unable to prove prior to the theory, and has provided students with a unified method for studying nonlinear programming and algorithmic convergence.

The text has three main sections. Chap. 1, which constitutes the first section, discusses how to formulate nonlinear programming problems. In it we treat production and inventory control, regression analysis, consumer behavior, chemical condenser design, rocket control, equation solving, cost-benefit analysis, and financial analysis as nonlinear programming problems. Included also are economic, business, governmental, mathematical, engineering, and scientific applications, as well as the topics of geometric programming, optimal control, and quadratic programming.

After formulating the problem, we begin to solve it by characterizing an optimal solution. The second portion of the book, Chaps. 2 and 3, delves into identifying a solution and treats such topics as concave and convex functions, the Kuhn-Tucker conditions, the constraint qualification, and duality theory. It also gives economic and business interpretations, as well as the dual geometric programming problem and the maximum principle of optimal control.

Armed with the knowledge of how to identify a solution point to a nonlinear programming problem, we attack a considerably more difficult problem, that of moving from a point that is not a solution to one that is. The study of this problem constitutes the third and final section of the book. Essentially, the problem leads to the study of algorithms, although we also discuss economic interpretations and present an important relationship between convergence theory and the Liapunov stability theory of difference equations.

The main thrust of the third section is embodied in the theory of convergence. Not only are several new algorithms presented and proved by means of convergence theory, but when previously known algorithms are considered we use the convergence theory instead of the old proofs. Thus a unified approach to convergence is established. Convergence theory provides a proof that is often considerably simpler than previously possible, although proving convergence is not without its difficulties. We hope, nevertheless, that the convergence theory will both reduce the student's effort and aid in providing convergence proofs to other algorithms.

Scope of the Book

No claim is made that this book is a compendium of all known results in nonlinear programming. However, the author has attempted to include a rich and varied selection of subjects that are of current interest. Many aspects of nonlinear programming not treated in the text are considered in the exercises. Thus, the exercises are an integral part of this book. In addition, the notes at the end of each chapter give references to source material and related topics, thereby pointing the way to even deeper investigation.

Prerequisites

The text assumes a certain knowledge of linear programming, advanced calculus, and linear algebra. Specifically, it presupposes an introduction to the linear programming problem, the simplex method, and the dual theorem of linear programming. The advanced calculus prerequisites include limits and sequences, continuity, partial differentiation, Taylor's expansion, and closed sets. An appendix is provided to refresh the reader's understanding of some of these topics. As for linear algebra, the text assumes the usual background in matrix and vector manipulation and linear-independence.

Even though certain other mathematical concepts are used in the text, they are fully explained before their use.

Course Presentation

The text contains sufficient material to serve either as the core of a year's course in nonlinear programming or as the basis for a few weeks' survey.

A short course could consist of the following chapters and sections: 2; 4; 5 (except 5.4.4); 7.1; 8.1; 8.2; 10.1; 11; 12.1; 12.2; 13.1; and 13.2. This presentation is logically consistent and covers many of the more important topics in nonlinear programming. Chapters 6 and 9 are more difficult and can be skipped on a first reading without any loss of continuity.

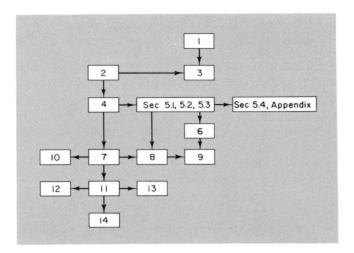

Fig. 1.

Introduction.

The logical interdependence of the chapters is depicted in Fig. 1. (Note that the book may be commenced with Chap. 2.)

Acknowledgments

I undertook and completed the book as a Ford Foundation Faculty Fellow and am indeed grateful to the Ford Foundation for its support. I am also grateful to Dr. Robert Lundegard, Mr. Marvin Denicoff, and the others in the Office of Naval Research, who, although they did not directly support the writing of this book, supported and encouraged my previous research, which was instrumental in providing the background for this book.

I am indebted to Mr. Sam Ginsburg who provided a detailed criticism of an early draft and proofreading assistance. The excellent example on consumer behavior in Chap. 1 is due to Mr. William Ziemba. Mr. Hans Tjian developed the lucid example of the convex-simplex-conjugate-direction technique. May I also express my deepest appreciation to Dr. Michael Canon of IBM, who was the first to rigorously obtain the geometric programming dual via the Kuhn-Tucker conditions. That section, in Chap. 3, is based on his notes.

For continued encouragement during this project, I thank Dean Richard Holton, Associate Dean Lee Preston, Professors C. West Churchman, Thomas Marschak, C. Bartlett McGuire, Romesh Saigal, B. Curtis Eaves, and many other of my colleagues at the School of Business Administration, University of California, Berkeley. In particular, Professor Eaves's detailed reading of the final draft saved me from many embarrassing misstatements.

To Professor Richard Cottle of Stanford goes my deepest appreciation for a detailed analysis of the final draft. Let me now single out Professor Arthur F. Veinott, Jr., of Stanford, who introduced me to the beauty and profundity of nonlinear programming. I am extremely grateful to him not only for his perceptive comments on the manuscript but more important for his continual encouragement over the years. Of course, I remain responsible for all errors and deficiencies.

Mrs. Julia Rubalcava, who typed the manuscript so well, and Mrs. Cordelia Thomas, whose editorial assistance was boundless, have earned my deepest gratitude. Finally, to my wife, Judith, and children, Michael and Monica, thank you for your good humor and help.

W. I. Z.

Berkeley, California

Table of Contents

Nonlinear

Programming

A Unified Approach

1 The Nonlinear Programming Problem

This chapter introduces the nonlinear programming (NLP) problem, presents several of its special forms, and illustrates it with detailed examples. The special forms include the geometric programming (GMP) problem, the optimal control (OC) problem, and the quadratic programming (QP) problem, while the examples illustrate its power to solve diverse and seemingly unrelated problems.

1.1. NONLINEAR PROGRAMMING

The NLP problem arises in a myriad of forms and may be found in the natural sciences, physical sciences, engineering, economics, mathematics, and in business and government. Indeed, even its philosophical implications have been analyzed. In its most abstract form, something must be maximized (or minimized). Suppose, for example, we must minimize the cost of both producing an item and holding it in stock by selecting an appropriate production schedule. However, the schedule cannot be chosen freely. Perhaps we must meet certain demand requirements for the item or there may also be limitations on the amount that can be produced at any one time. Thus we must select the production schedule that minimizes cost, yet that also satisfies certain other constraints.

Generally, for the NLP problem something must be maximized (or minimized); however, constraints limit the actions we may take to achieve this maximum (or minimum).

In the mathematical context we are given a function $f(x)$. The point x is a point in E^n, where E^n denotes n-dimensional Euclidean space (x is an n-dimensional column vector), and $f(x)$ is a point in E^1. Our goal is to select an x to maximize $f(x)$. For this reason f is termed the **objective function**. However, the point x may not be selected arbitrarily because we are given m **constraint** functions, $g_i(x)$, $i = 1, \ldots, m$, and the x selected must satisfy

$$g_i(x) \geq 0 \qquad i = 1, \ldots, m$$

Consequently, we attempt to maximize $f(x)$ over the set delimited by the above constraints. A point x^* that achieves the maximum is called **optimal**.

Standard Form

To state the NLP problem in standard form, some notation is needed. The phrase

$$h \quad \text{is a function} \quad h: E^n \longrightarrow E^1$$

means that h acts upon any point x in E^n and yields a point $h(x)$ in E^1.
The standard form of the NLP problem seeks an optimal point x^* to

$$\max \quad f(x)$$
$$\text{subject to} \quad g_i(x) \geq 0 \qquad i = 1, \ldots, m \tag{1.1}$$

where the functions f and g_i are $f: E^n \to E^1$ and $g_i: E^n \to E^1, i = 1, \ldots, m$.
The function f is the objective function, and the g_i are constraints. Any x
that satisfies all the constraints, that is any x such that

$$g_i(x) \geq 0 \qquad i = 1, \ldots, m$$

is termed **feasible**. We call the set of all feasible points the feasible set and
denote it by F. To avoid unnecessary complications assume $F \neq \phi$, where
ϕ is the null set.

Abstract Example

As a brief example consider the following problem where $x = (x_1, x_2, x_3, x_4)^t$ and the superscript t indicates transpose.

$$\min \quad f(x) \equiv (x_1)^2 + 4x_1 e^{x_2} + x_2 x_3 (x_4)^2$$

$$\text{subject to} \quad g_1(x) \equiv 3 + x_1 + x_2 + x_3 + \frac{3}{x_4} \geq 0$$

$$g_2(x) \equiv 1 + 2x_1 - x_2 + (x_3)^3 + 4e^{x_4} \leq 100$$

$$g_3(x) \equiv 2 + 4x_1 + 2x_2 + x_4 = 0$$

$$g_4(x) \equiv x_1 \geq 0$$

$$g_5(x) \equiv x_2 \geq 0$$

$$g_6(x) \equiv x_3 \geq 0$$

$$g_7(x) \equiv x_4 \geq 1$$

This problem has a nonlinear objective function and two nonlinear
constraints. Observe that the goal is to minimize rather than maximize as in
the standard form. The maximizing and minimizing formulations are equivalent, as multiplying the objective function by -1 transforms a minimization
to a maximization. Note also the equality for g_3 and the reversed inequality
for g_2 (see exercise 1.1).

1.2. USES OF NONLINEAR PROGRAMMING

Economic, Business, and Government Applications

The main objective of NLP and, in fact, of this book is to specify the means of calculating the optimal point for problem (1.1), the NLP problem. However, NLP also presents a powerful conceptual framework for problem formulation, even if the various functions involved cannot be explicitly determined or if it is impossible to precisely calculate the optimal point.

For example, the NLP problem is intimately related to the basic economic problem. Economics has been said to be the science of allocating scarce resources in a manner either to maximize efficiency or, if one is dealing with consumers, to maximize consumer utility. NLP also fits this framework. The objective function can indicate efficiency, which we attempt to maximize, while the constraints can specify the limitations imposed by the scarce resources. Analogously, the objective function can be consumer utility. Thus we see the direct connection between the NLP problem and the basic economic problem.

In such a general setting, determining the precise form of the functions may be impossible; nevertheless, in specific applications precise formulation is often straight forward. Consider a particular industrial plant, say a plastics manufacturer. Here, efficiency may become profit, and the constraints may be interpreted as manpower, space available, machine capacities, etc. In such a specific situation the quantitative data are often available, and the NLP problem can then be precisely formulated and solved.

Cost-Benefit Analysis

Also within the NLP context is the important concept of cost-benefit analysis. Cost-benefit analysis was developed to assist decision making in government where there is no profit function but instead there is a general welfare-benefit function. Two intimately related NLP problems arise here, either maximize benefit subject to a cost limitation or minimize cost with the benefits above a minimum level (see exercise 1.2). With the vast amount of data available to various government agencies, particular cost-benefit analyses can often be well modeled by NLP. Even when the problem is too elusive for explicit NLP formulation, first approximations are often available through NLP or various portions of the problem are solvable through NLP.

The aforementioned applications of NLP have concentrated on decision making problems. Indeed, aiding an individual, executive, or governmental decision maker is a NLP forte. Of course, the astute decision maker does not accept outright that the optimal point for the NLP problem is the best answer for his real-world problem. The optimal point is only advisory, and the decision maker should investigate the assumptions and accuracy of the NLP formulation before making his decision. Nevertheless, numerous programming problems have proven themselves in practice, and in these cases the decision implied by the optimal point is implemented almost without question.

Solving Equations

An important application of NLP to the mathematical, physical, and social sciences arises because NLP is perhaps the most powerful general technique for solving systems of equations. To illustrate this, suppose we must solve the system of equations

$$a_i(x) = 0 \qquad i = 1, \ldots, l$$

Let us form the objective function

$$f(x) = \sum_{i=1}^{l} (a_i(x))^2$$

Certainly

$$f(x) \geq 0$$

and

$$f(x) = 0 \quad \text{if and only if } x \text{ solves the equations}$$

Solving the NLP problem of minimizing f subject to no constraints will, in fact, solve the equations (see exercise 1.3).

Scientific Applications

Explicit NLP problems often arise in the sciences. In physics, for example, the objective function might be potential energy and the constraints the various equations of motion. Minimizing the objective function would determine a stable configuration of the system. Correspondingly, by changing the objective function, we can determine the configuration with the largest thermal energy, kinetic energy, etc. A related problem in chemistry is to determine the molecular structure that minimizes the Gibbs free energy. For the social and psychological sciences the problem arises of minimizing social tension where individuals or groups are constrained to follow certain behavior patterns.

A Caution

The actual transformation of a real problem into a NLP problem is largely an art; yet it is an art tempered by theory. Primarily the theory forms a framework that facilitates problem formulation. But equally as important the theory pinpoints which of the many possible formulations is most efficiently solved and which cannot be solved at all. Indeed, as the theory will reveal, we may find it almost impossible, in practice, to calculate the optimal point, and we may have to settle for much less. Most of this book is devoted to a presentation of this theory.

1.3. A DETAILED EXAMPLE: A REGRESSION PROBLEM IN CONSUMER CHOICE

Suppose in a given month a consumer allots E dollars to spend on k different products. If he purchases q_i of product i at a price p_i, then

$$\sum_{i=1}^{k} p_i q_i = E \tag{1.2}$$

However, he does not allocate his budget haphazardly, for every month he purchases a certain minimal amount $\bar{q}_i$ of product i. Therefore he spends at least

$$\sum_{i=1}^{k} p_i \bar{q}_i$$

on the k products. Out of the remaining money a certain fixed fraction a_i of funds is allocated to purchase more of product i. The a_i may be interpreted as the marginal propensity to consume commodity i out of uncommitted funds.

The total number of dollars spent on product i in a month is then

$$p_i q_i = p_i \bar{q}_i + a_i \left[E - \sum_{j=1}^{k} p_j \bar{q}_j \right] \tag{1.3}$$

Here $p_i \bar{q}_i$ is the cost of the minimal purchase of product i while

$$a_i \left[E - \sum_{j=1}^{k} p_j \bar{q}_j \right]$$

is the additional funds allotted to product i. Furthermore, since the a_i are fractions

$$\sum_{i=1}^{k} a_i = 1 \tag{1.4}$$

$$a_i \geq 0$$

Let us now suppose that a given individual purchases according to the above model. That is, each month he purchases roughly $\bar{q}_i$ basic amount of product i and after his basic needs are satisfied he allots approximately

$$a_i \left[E - \sum_{j=1}^{k} p_j \bar{q}_j \right]$$

amount of funds to purchase additional quantities of product i. Our problem is to infer the numbers $\bar{q}_i$ and a_i from his observed purchase behavior over several months. Once these numbers can be estimated, we have a model to predict the consumer's behavior.

Let $E_l, l = 1, \ldots, L$, be the total amount spent in month l. Also suppose that he purchased a total amount q_{il} of product i at a price p_{il} in month l. Then according to the model, from Equation (1.3), we find that

$$p_{il} q_{il} = p_{il} \bar{q}_i + a_i \left[E_l - \sum_{j=1}^{k} p_{jl} \bar{q}_j \right] + \epsilon_{il} \tag{1.5}$$

where ϵ_{il} is the residual or error. Here the $\bar{q}_i$ and a_i, because they reflect the consumer's personality, are assumed constant over the several months considered. However, the budget E_l, prices p_{il}, and quantities purchased q_{il} differ in each month.

In order to find good estimates of $\bar{q}_i$ and a_i the following NLP problem could be solved. Recall, $\bar{q}_i$ and a_i are the variables to be determined given the known numbers E_l, p_{il}, and q_{il}. The ϵ_{il} are also variables.

$$\min \sum_{i=1}^{k} \sum_{l=1}^{L} (\epsilon_{il})^2$$

$$\text{subject to}\quad p_{il}q_{il} - p_{il}\bar{q}_i - a_i\left[E_l - \sum_{j=1}^{k} p_{jl}\bar{q}_j\right] = \epsilon_{il} \qquad \begin{aligned} i &= 1, \ldots, k \quad \text{and}\\ l &= 1, \ldots, L \end{aligned}$$

$$\sum_{i=1}^{k} a_i = 1$$

$$a_i \geq 0 \qquad i = 1, \ldots, k$$

$$q_{il} \geq \bar{q}_i \qquad \begin{aligned} l &= 1, \ldots, L \quad \text{and}\\ i &= 1, \ldots, k \end{aligned}$$

$$\bar{q}_i \geq 0 \qquad i = 1, \ldots, k$$

The objective function minimizes squared error. The first constraint is simply Equation (1.5), while the second and third constraints ensure that the a_i are the fractions required by (1.4). The final two constraints provide for the consistency of the $\bar{q}_i$.

1.4. GEOMETRIC PROGRAMMING

Geometric programming (GMP) is a special form of NLP that has been found particularly useful in engineering and the physical sciences. To introduce the topic consider the following example.

Example

As vice president of a small but dynamic chemical company you have been called upon to investigate containerization for shipping. Not only should the containers simplify shipment, but also because the containers will be left with the customers, sales should be enhanced. You decide to examine a particular problem, specifically, each month 1,000 cubic feet of a chemical must be shipped to a customer. The chemical is to be sent in rectangular containers of length w_1, width w_2, and height w_3. The sides and bottom must be made of scrap material for which there is no charge. However, only 10 square feet of which becomes available per container each month. Material for the ends costs \$2 per square foot while material for the top costs \$3 per square foot. There is also a shipping charge of 20 cents per container sent. You must determine how many and what size of containers are needed to ship the chemical at the lowest possible cost.

Minimizing the total cost (shipping plus the costs of the two ends and the top for 1,000 $w_1 w_2 w_3$ containers) subject to the scrap limitation

$$\text{min}\quad 0.2\left(\frac{1,000}{w_1w_2w_3}\right) + 4w_2w_3\left(\frac{1,000}{w_1w_2w_3}\right) + 3w_1w_2\left(\frac{1,000}{w_1w_2w_3}\right)$$

$$\text{subject to}\quad 2w_1w_3 + w_1w_2 \leq 10$$

$$w_1 > 0 \qquad w_2 > 0 \qquad w_3 > 0$$

Letting

$$h_0(w) = \frac{200}{w_1w_2w_3} + \frac{4,000}{w_1} + \frac{3,000}{w_3}$$

and

$$h_1(w) = \frac{w_1w_3}{5} + \frac{w_1w_2}{10}$$

where $w = (w_1, w_2, w_3)^t$, we may rewrite the problem as

$$\text{min}\quad h_0(w)$$

$$\text{subject to}\quad h_1(w) \leq 1 \qquad w > 0$$

Recall superscript t indicates transpose.

General Formulation

We now develop the general mathematical form of the GMP problem. Key to this development is the **posynomial function**. A function $h(w)$ where $w = (w_1, \ldots, w_n)^t$ is a posynomial if h can be expressed as the sum of power terms each of the form

$$c_i w_1^{a_{i1}} w_2^{a_{i2}} w_3^{a_{i3}} \cdots w_n^{a_{in}}$$

where the a_{ij} and c_i are constants

$$c_i > 0 \quad \text{and} \quad w_j > 0$$

The requirement that c_i be positive gives rise to the syllable *posy* in posynomial, while w_i positive ensures that $w_j^{a_{ij}}$ is defined for a_{ij} fractional.

Thus h is a posynomial if

$$h(w) = \sum_{i=1}^{k} c_i \left[\prod_{j=1}^{n} w_j^{a_{ij}} \right]$$

where $c_i > 0$ and $w_j > 0$. Here Π denotes product.

The Geometric Programming Problem

We may now pose the GMP problem:
Given $p + 1$ posynomials $h_0, \ldots, h_p$ determine w to

$$\begin{aligned} \min \quad & h_0(w) \\ \text{subject to} \quad & h_k(w) \leq 1 \quad k = 1, \ldots, p \\ & w > 0 \end{aligned} \qquad (1.6)$$

It is left as an exercise (see Exercise 1.8) to show that the GMP problem may be reformulated after revising the functions as

$$\begin{aligned} \min \quad & Ln\, h_0(y) \\ \text{subject to} \quad & Ln\, h_k(y) \leq 0 \quad k = 1, \ldots, p \\ & y - Ax = b \end{aligned} \qquad (1.7)$$

and

$$h_k(y) = \sum_{i=n_k}^{m_k} e^{y_i} \quad k = 0, 1, \ldots, p$$

where $y = (y_1, y_2, \ldots, y_{m_p})^t$ and $x = (x_1, \ldots, x_n)^t$ are variables, and the vector $b = (b_1, \ldots, b_{m_p})^t$ and $m_p \times n$ matrix A are given constants. Also $n_0 = 1$, $n_1 = m_0 + 1$, $n_2 = m_1 + 1, \ldots, n_p = m_{p-1} + 1$. The notation Ln, of course, denotes the logarithm to the base e.

In Chap. 3 we will demonstrate how the rather formidable appearing GMP problem can be greatly simplified.

1.5. THE OPTIMAL CONTROL PROBLEM

Optimal control is used in numerous dynamic engineering and economic problems requiring the selection of control. Certain control variables govern the evolution of the system from one period to the next and must be selected to maximize (or minimize) an objective function.

Two types of variables are essential to the problem—the control variables, $u^i = (u^i_1, \ldots, u^i_m)^t$, which have already been mentioned, and the state variables, $x^i = (x^i_1, \ldots, x^i_n)^t$, which entirely describe the system at time i.

Given the state x^i at time i, we specify the change in the state from time i to $i + 1$ by a vector transformation function $g^i(x^i, u^i) = (g^i_1, \ldots, g^i_n)^t$. Thus in vector notation the evolution of the system from time i to $i + 1$ is

$$x^{i+1} = x^i + g^i(x^i, u^i)$$

Note that the change in state depends upon both the state and the control.

A profit (or cost) $f^i(x^i, u^i)$ is also associated with being in state x^i and using control u^i.

To complete formulation of the problem the initial state of the system x^1 must be specified. Also the final state of the system, x^{l+1}, may be given either explicitly or implicitly by requiring x^{l+1} to satisfy certain equations

$$q(x^{l+1}) = 0$$

where

$$q = (q_1, \ldots, q_s)^t$$

The Optimal Control Problem

We may now formulate the OC problem as

$$\max \quad \sum_{i=1}^{l} f^i(x^i, u^i)$$

$$\text{subject to} \quad g^i(x^i, u^i) + x^i = x^{i+1} \qquad i = 1, \ldots, l \tag{1.8}$$

$$q(x^{l+1}) = 0$$

where x^1 is given. (Recall, the g^i and q are vectors.)

Explicitly, the OC problem, by selecting appropriate control variables, seeks to maximize the total profit over the l periods. However, the system must satisfy a certain initial condition, x^1, evolve in a specified manner, $g^i(x^i, u^i) + x^i = x^{i+1}$, and result in a given terminal condition, $g(x^{l+1}) = 0$.

For certain more general OC problems the controls may be constrained

$$h^i(u^i) \geq 0 \quad \text{where} \quad h^i = (h^i_1, \ldots, h^i_{k_i})$$

Also it is possible that only certain states are permitted so that the state variables are themselves constrained

$$p^i(x^i) \geq 0 \quad \text{where} \quad p^i = (p^i_1, \ldots, p^i_{e_i})$$

The OC problem will be further explored in Chap. 3 where we will develop the *maximum principle* of OC. However, to conclude this section consider the following example.

Example

An unmanned rocket in outer space must travel from one fuel depot to another depot, a distance of 1 million miles. Assume that no external forces operate on the rocket and that the rocket travels in a frictionless straight-line path. At discrete points every 100,000 miles the rocket's thrust may be instantaneously (for all practical purposes) adjusted, but between these control points no thrust changes are possible. Although the total amount of thrust is limited, it is desired to traverse the million miles in minimum time. How should the thrust of the rocket be controlled?

Let v_i be the rocket's instantaneous speed in miles per hour as it passes the ith control point. Because the rocket starts at the initial control point and stops at the 11th point,

$$v_1 = 0$$

and

$$v_{11} = 0$$

Let u_i be the constant thrust as it travels from control point i to $i + 1$. If m is the mass of the rocket in appropriate units and a_i the constant acceleration between points i and $i + 1$, then Newton's second law requires

$$u_i = ma_i \tag{1.9}$$

Since the distance between check points is 100,000 miles, the laws of motion require

$$100{,}000 = v_i t_i + \tfrac{1}{2} a_i (t_i)^2 \tag{1.10}$$

where t_i is the time taken to travel from control point i to $i + 1$. Also

$$v_{i+1} = v_i + a_i t_i \tag{1.11}$$

via (1.10) since time is positive and assuming $a_i \neq 0$

$$t_i = \frac{-v_i + [(v_i)^2 + 200{,}000 a_i]^{1/2}}{a_i} \tag{1.12}$$

From Equation (1.12), Equation (1.11) becomes

$$v_{i+1} = v_i + a_i \left\{ \frac{-v_i + [(v_i)^2 + 200{,}000 a_i]^{1/2}}{a_i} \right\}$$

Simplifying and using Equation (1.9)

$$v_{i+1} = \left[(v_i)^2 + 200{,}000 \frac{u_i}{m} \right]^{1/2}$$

The OC problem of minimizing total time then becomes

$$\min \quad \sum_{i=1}^{10} \left\{ \frac{-v_i + [(v_i)^2 + 200{,}000u_i/m]^{1/2}}{u_i/m} \right\}$$

$$\text{subject to} \quad v_{i+1} = \left[(v_i)^2 + 200{,}000 \frac{u_i}{m} \right]^{1/2} \qquad i = 1, \ldots, 10$$

$$-b_i \leq u_i \leq c_i \qquad v_1 = v_{11} = 0$$

where c_i is the maximum thrust allowable in the forward and b_i in the backward direction between points i and $i + 1$.

1.6. QUADRATIC PROGRAMMING PROBLEM

The QP problem is simply the NLP problem with a quadratic objective function and linear constraints. It is usually formulated as follows

$$\max \quad q^t x + \tfrac{1}{2} x^t Q x \tag{1.13}$$
$$\text{subject to} \quad Ax = b \qquad x \geq 0$$

where q is an n vector, Q is a symmetric $n \times n$ matrix, b is an m vector, and A is an $m \times n$ matrix.

Of all the various NLP problems the QP problem is generally the easiest to solve. Indeed, it is only slightly more complex than the linear programming (LP) problem.

Example

A financier is debating how to allot his funds among n possible investments. Suppose investment i has an expected profit of $\bar{p}_i$ per dollar invested in it. Then if x_i is the amount invested in the ith investment, the expected profit becomes

$$\sum_{i=1}^{n} \bar{p}_i x_i$$

In addition, the investment portfolio $x = (x_i)$ must satisfy certain constraints

$$Ax = b$$
$$x \geq 0$$

where A is an $m \times n$ matrix and b is an m vector. These mathematical constraints reflect the financier's actual constraints such as total funds, limits on the amount of funds placed in any one investment category, etc.

A Linear Programming Approach

A first approach to the financier's problem would be to solve the LP problem of maximizing expected profit subject to the constraints.

$$\max \quad \sum_{i=1}^{n} \bar{p}_i x_i$$

$$\text{subject to} \quad Ax = b \qquad x \geq 0$$

A Quadratic Programming Formulation

However, it might be that the profit has a rather large variance. The above linear programming model does not consider variance and could lead to a poor investment decision.

To incorporate variance in our analysis, let q_{ij} be the covariance between investment i and j per unit of dollar invested in each. Then the covariance matrix becomes

$$Q = (q_{ij})$$

and the variance of any given portfolio x is

$$x^t Q x$$

Another formulation of the financier's problem would be to select the portfolio that minimizes variance yet achieves an expected profit of at least some fixed amount c. We are thus led to the QP problem

$$\min \quad x^t Q x$$

$$\text{subject to} \quad Ax = b$$

$$\sum_{i=1}^{n} \bar{p}_i x_i \geq c \qquad x \geq 0$$

1.1. The basic NLP problem was formulated with only inequality constraints.

(a) Show that an equality constraint can be represented by two inequality constraints.

(b) Show how to represent an inequality constraint by an equality constraint.

Hint: $g(x) - y^2 = 0$.

1.2. The cost-benefit problem can be formulated as

$$\max \quad b(x)$$
$$\text{subject to} \quad c(x) \leq C$$

where $b(x)$ is the benefit of allocation x, $c(x)$ is the cost of this allocation, and C is the cost limitation. Suppose the optimal solution to the above problem is x^*. Under what conditions would x^* also be the solution to the problem

$$\min \quad c(x)$$
$$\text{subject to} \quad b(x) \geq b(x^*)$$

Interpret this problem economically.

1.3. Consider the problem of solving the system of equations and inequalities

$$a_i(x) \geq 0 \qquad i = 1, \ldots, l$$
$$a_i(x) = 0 \qquad i = l+1, \ldots, L$$

(a) Formulate this problem as a NLP problem without constraints.

(b) Formulate this problem as a NLP problem with constraints.

(c) Under what circumstances would (a) be preferable to (b)?

1.4. *Production Planning.* Let x_i be the amount instantaneously produced in the beginning of period i and I_i be the amount of stock held at the end of period i, $i = 1, \ldots, n$. Suppose during the middle of period i a known demand r_i for the product occurs that must be satisfied immediately. There is a cost of producing $p_i(x_i)$ and a cost of holding inventory $h_i(I_i)$. Also production in each period is limited by a capacity constraint c, while the inventory level must be less than d. The initial inventory is zero.

Formulate the NLP problem of minimizing total cost. Show that the problem is also an OC problem. What are the control variables, and what are the state variables?

1.5. *Nonlinear Constrained Regression Analysis.* Suppose l observations of the variables $x, z_1, z_2, \ldots, z_m$ have been made according to the model

$$x_i = h(z_{1i}, z_{2i}, \ldots, z_{mi}, \alpha_1, \alpha_2, \ldots, \alpha_p) + \epsilon_i \qquad i = 1, \ldots, l$$

in which $x_i, z_{1i}, \ldots, z_{mi}$ indicate the values taken by the variables at observation i, $i = 1, \ldots, l$, the $\alpha_1, \ldots, \alpha_p$ are p unknown parameters to be estimated, and ϵ_i is an unobservable random variable with zero mean. The function h is a known function of its $m + p$ arguments.

Suppose in addition that the model requires the α_i to satisfy the following conditions:

$$g_j(z_{1i}, z_{2i}, \ldots, z_{mi}, \alpha_1, \alpha_2, \ldots, \alpha_p) \geq 0$$

for $j = 1, \ldots, k$ and $i = 1, \ldots, l$.

Indicate how one could use NLP to determine good estimates of the α_i.

1.6. *Inventory Control: Economic Lot Size.* Let r be the fixed demand rate per unit time for a product. When just out of stock we order an amount Q which is delivered instantaneously. Consequently an order for Q is placed every Q/r periods, and the average inventory level is $Q/2$.

Suppose that the cost of ordering Q is $K + cQ + dQ^2$ where $K > 0$, $c > 0$, and $d > 0$, and that the inventory holding cost is h per unit held in stock per unit time.

Determine Q to minimize cost per period if it is known that Q cannot be more than an amount q. Show that this is a GMP problem. Solve the problem. Interpret K, c, d, and q.

1.7. *Vapor Condenser Design.* After various simplifying assumptions the problem of designing a horizontal vapor condenser may be specified as follows: Let N be the number of tubes, D the average tube diameter in inches, and L the tube length in feet. Then for certain constants, $\beta_1, \beta_2, \beta_3, \beta_4$, thermal energy is

$$\frac{\beta_1}{N^{7/6} D L^{4/3}} + \frac{\beta_2}{D^{0.2} L}$$

fixed charges on the heat exchange are

$$\beta_3 N D L$$

and the cost of pumping cold fluid is

$$\frac{\beta_4 L}{D^{4.8} N^{1.8}}$$

We wish to minimize the fixed charges on heat exchange plus the cost of pumping cold fluid with the constraint that thermal energy be less than a fixed amount c.

In addition, the condenser is square in cross section and to fit can have a width and height of at most w inches. Each tube occupies a cross sectional square of width and height D, its diameter.

Formulate the GMP problem of minimizing cost subject to the thermal energy and cross-sectional size limitations.

1.8. (a) Prove that problem (1.6) is the same as problem (1.7).
  Hint: Let $w = e^x$. Observe $e^x > 0$.
  (b) When does the GMP problem reduce to a LP problem?

1.9. Under what realistic conditions might the programming problems previously given become QP problems?

1.10. *Geometric Interpretation of NLP.* Graph the two feasible sets

(a)
$$(x_1)^2 + (x_2)^2 \leq 4$$
$$x_1 \geq 0 \qquad x_2 \geq 0$$

(b)
$$e^{x_1 + x_2} \geq 4$$
$$x_1 + x_2 \leq 10$$
$$x_1 \geq 0 \qquad x_2 \geq 0$$

1.11. Depict the objective function $x_1 + 2(x_2)^2$ subject first to the constraints in 1.10(a) above and second to (b). From these pictures give reasonable solutions to the two problems of maximizing this objective function over each of the feasible sets.

1.12. *International Trade* (KARLIN). Suppose w_i, x_i, and y_i are respectively the amount of commodity i imported, exported, and produced domestically. Let a_{ij} be the amount of commodity i used in producing one unit of product j, l_i the labor per unit of product i produced, and c_i the capital required per unit of product i produced. Also suppose it costs g_i per unit to import product i while a revenue of h_i is obtained per unit of product i exported. The revenue h_i can be expressed as $h_i = \gamma_i + \rho_i x_i$ where $\gamma_i > 0$ and $\rho_i < 0$. Finally, let d_i be the total demand for product i.

We desire to find the production schedule that minimizes total capital required yet that also satisfies demand. However, total labor is limited to an amount L, while the trade deficit cannot exceed D. By trade deficit is meant the total cost of imports less the values of exports. Formulate the NLP problem.

1.13. The per unit production of product i uses a_{ij} amount of resource j, but only $b_j > 0$ units of resource j are available. Let c_i, a constant, be the selling price of product i, and suppose the cost of producing the xth unit of product i is $u_i(x)$. What is the total cost of producing x units of product i? Write the programming problem of maximizing net profit. What conditions would the u_i have to satisfy in order for the problem to be a QP problem? Interpret these conditions.

1.14. *Chemical-Equilibrium; Gibbs Free Energy* (DANTZIG, 1963). We are interested in calculating the number of moles of n different types of gas molecules present in a gaseous mixture at equilibrium. Let $x_j, j = 1, \ldots, n$, be the number of moles of molecular species j present. Certainly $x_j \geq 0$. Then the Gibbs free energy of the mixture is

$$G(x) = \sum_{j=1}^{n} c_j x_j + \sum_{j=1}^{n} x_j \log \left(\frac{x_j}{\bar{x}} \right)$$

where $\bar{x} = \sum_{j=1}^{n} x_j$, and the c_j are given parameters called Gibbs free energy functions which depend upon the temperature, pressure, the molecular species, and the universal gas constant.

Suppose m atoms comprise the mixture and that b_i, $i = 1, \ldots, m$, is the number of atomic weights of atom i present. If a_{ij} is the number of atoms of type i in a molecule of species j, then the mass balance equations stipulate

$$\sum_{j=1}^{n} x_j a_{ij} = b_i \qquad i = 1, 2, \ldots, m$$

The equilibrium mixture can be determined by minimizing the Gibbs free energy subject to the mass balance restrictions and the $x_j \geq 0$. Formulate the corresponding NLP problem.

NOTES AND REFERENCES

§1.1. The NLP problem has been treated extensively. For introductory discussions see the works of WAGNER and of HILLIER and LIEBERMAN. Various review articles containing some of the highlights up to the time of their publication are DORN (1963); SPANG; ZOUTEDNIJK (1966). For a practical slant see HADLEY (1960). A more classical approach to optimization can be found in the work of HANCOCK. The books of ABADIE and of GRAVES and WOLFE are collections of scholarly articles by notables in the field. A book by DENNIS discusses certain aspects of NLP from an electrical engineer's viewpoint. Other works include BERGE and GHOUILA-HOURI; HADLEY (1964); KARLIN; KUNZI and KRELLE; SAATY and BRAM; VAJDA; WILDE and BEIGHTLER.

§1.2. NLP has been applied to foreign trade and exchange by REITER. The problem of blending octane numbers treated from a NLP viewpoint is examined by MANNE (1956). Aspects of NLP flourish in consumer buying theory; see SAMUELSON (1938).

The classic work in cost-benefit analysis and its twin brother program planning and budgeting is that of HITCH and MCKEAN. For the chemical theory of Gibbs free energy see GIBBS. The problem of treating Gibbs free energy as a NLP problem is often referred to as the chemical-equilibrium problem; see works by DANTZIG, JOHNSON, and WHITE; BRINKLEY; WHITE, JOHNSON and DANTZIG; DUFFIN, PETERSON, and ZENER (see also exercise 1.14).

§1.3. This example is due to W. ZIEMBA. For further discussion of this problem see MALINVAUD, who treats it from a more classical viewpoint, and STONE, who examines data. A general treatment of utility theory can be found in DEBREU (1954).

§1.4. The example is an adaptation of an example in DUFFIN, PETERSON and ZENER. (See also that book for a thorough treatment of GMP, including numerous examples such as transformer design and the conversion of solar energy into useful energy by extracting heat from ocean water.) An interesting exposition of GMP is also contained in WILDE and BEIGHTLER.

§1.5. Optimal control is an extensive topic to which this short section cannot do justice. The classic work is PONTRYAGIN *et al.* An excellent treatment can also be found in HESTENES. For an analysis of the relation between OC and NLP, see CANON, CULLUM, and POLAK. Applications of OC are presented in LEITMAN; CONNORS and TEICHROEW.

§1.6. The example is based upon the work of MARKOWITZ. Quadratic programming will be treated further in Chap. 8. An exposition of the topic is contained in the books by BOOT (1964); KUNZI and KRELLE.

Exercises: **1.5.** For further treatment of regression, see MALINVAUD.

1.7. This problem is adapted from AVRIEL and WILDE; see also WILDE and BEIGHTLER.

ADDITIONAL COMMENTS

It is impossible in this text to treat all aspects of NLP and also all topics related to NLP. For example, integer programming, in which certain variables are constrained to be integral, is a topic unto itself. A basic work in this area is that of GOMORY. More recent discussions can be found in BALINSKY; Chap. 9 of ABADIE; GLOVER.

A second area that will not be examined is stochastic programming, which assumes some or all of the parameters are random. For an approach called programming under uncertainty, which specifies a loss in utility or penalty for uncertainty, see DANTZIG (1955); or WETS. Chance constrained programming [see CHARNES and COOPER (1959)] emphasizes safety factors. The approach of TINTER is useful for economic planning. An interesting relation between programming under uncertainty and OC is found in VAN SLYKE and WETS. They also have many references. A Markov chain method can be found in BLACKWELL; D'EPENOUX; HOWARD; MANNE (1960).

The topic of dynamic programming is also too enormous to undertake in this text. Dynamic programming is a general optimization theory for problems that can be transformed into a sequence of stages. Nearly all optimization problems can be so transformed. For introductory treatments of this topic, see HILLIER and LIEBERMAN; WAGNER; NEMHAUSER. The basic works in the field remain BELLMAN (1957); BELLMAN (1961); BELLMAN and DREYFUS.

2 Identifying

an Optimal Point

The previous chapter was devoted to formulating the NLP problem, which we must now solve. Before a problem can be solved, the optimal point must be identified. Recall a point x^* is optimal if it maximizes the objective function subject to the constraints. This chapter studies criteria that facilitate the identification of optimal points.

The chapter begins with a vector that is of immense importance in nonlinear programming (NLP), the gradient vector. We use it both to develop the necessary conditions for a NLP problem without constraints and to develop the Kuhn-Tucker (K-T) necessary conditions for the constrained problem. The sufficient conditions that a point be optimal will be developed by two approaches, via the concave function and its variants, and via duality theory.

2.1. THE GRADIENT VECTOR

The importance of the gradient vector to NLP should not be understated; for given a nonoptimal point, the gradient usually aids in obtaining a better point. However, before discussing the gradient, we must develop the concept of a direction, because the gradient is itself a direction.

A Direction Vector

Any n-dimensional column vector d may serve as a direction. Suppose a point $x \in E^n$ and direction d, $d \neq 0$, are given. As we vary the scalar τ between 0 and $+\infty$, the point y where

$$y = x + \tau d$$

describes a ray emanating from the point x in the direction d. By varying τ from $-\infty$ to $+\infty$, y determines an entire line through x. The direction $d = 0$, where 0 denotes the zero vector, satisfies the mathematical definition of direction as it is an n-dimensional vector, although the zero direction is best regarded as a mathematical convenience (see exercise 2.1).

The Gradient

For any differentiable function f, the gradient at x is the vector of partial derivatives evaluated at x

$$\nabla f(x) = \begin{pmatrix} \dfrac{\partial f(x)}{\partial x_1} \\ \cdot \\ \cdot \\ \cdot \\ \dfrac{\partial f(x)}{\partial x_n} \end{pmatrix}$$

Not only is the gradient a direction but it is an extremely important one. In particular, given an arbitrary direction d, $\nabla f(x)^t d$ describes the instantaneous rate of change of f along the direction d. More precisely it is proved in advanced calculus that if f is differentiable at x

$$\lim_{\tau \to 0} \frac{f(x + \tau d) - f(x)}{\tau} = \nabla f(x)^t d \qquad (2.1)$$

The next theorem reviews what occurs should this instantaneous rate of change be positive. Recall the superscript t indicates transpose.

Theorem 2.1: *Let f be differentiable at x. Suppose there is a direction d such that*

$$\nabla f(x)^t d > 0$$

Then a $\sigma > 0$ exists such that for all τ, $\sigma \geq \tau > 0$

$$f(x + \tau d) > f(x)$$

PROOF: Via Equation (2.1)

$$\lim_{\tau \to 0} \frac{f(x + \tau d) - f(x)}{\tau} > 0 \qquad (2.2)$$

The definition of limit provides that there must be a $\sigma > 0$ such that for all $\tau \neq 0$ and $\sigma > \tau > -\sigma$

$$\frac{f(x + \tau d) - f(x)}{\tau} > 0 \qquad (2.3)$$

Select $\tau > 0$ to preserve the inequality and the result holds.   ◆

Intuitively, the gradient, if it is not zero, points in a direction such that a small movement in that direction will increase f. Moreover, suppose we are given any direction d that points in a direction similar to the gradient in the sense that $\nabla f(x)^t d > 0$. Then, as proved in Theorem 2.1, a small movement in that direction will also increase f. Theorem 2.1 is quite important; roughly it states that the gradient points "up hill."

2.2. NECESSARY OPTIMALITY CONDITIONS
FOR THE UNCONSTRAINED PROBLEM

The NLP problem is said to be unconstrained if it has no constraints. An immediate corollary of Theorem 2.1 provides a necessary condition that x^* be optimal for an unconstrained problem. Specifically, the condition is that

$$\nabla f(x^*) = 0$$

Corollary 2.1.1: *Let f be differentiable. If x^* maximizes f over E^n, then*

$$\nabla f(x^*) = 0$$

PROOF: Suppose $\nabla f(x^*) \neq 0$. Selecting $d = \nabla f(x^*)$

$$\nabla f(x^*)^t d = \nabla f(x^*)^t \nabla f(x^*) > 0$$

Then by Theorem 2.1 there would be a point higher than $f(x^*)$. The contradiction is evident. ◆

Although the condition $\nabla f(x^*) = 0$ necessarily follows should x^* be an unconstrained maximum, the condition is certainly not sufficient. In fact, even if $\nabla f(x^*) = 0$, x^* might be a local maximum, a global minimum, or even a saddle point (see exercises 2.3 and 2.4). At a local maximum x^* maximizes f over a neighborhood of x^*, while at a global maximum x^* maximizes f over all of E^n. The term saddle point is defined in Sec. 2.6.

The next section describes functions for which the necessary condition that x^* be a global maximum, viz., $\nabla f(x^*) = 0$, is also sufficient to guarantee that x^* is a global maximum.

2.3. CONCAVITY, CONVEXITY, PSEUDOCON-
CAVITY, AND QUASI-CONCAVITY

Concave functions are especially valuable in NLP. They, as will be seen, ensure that the necessary conditions for x^* to be optimal in either the unconstrained or the constrained case are also sufficient.

Convex Sets

Because concave functions are usually defined on and intuitively related to convex sets we investigate the latter concept first. Consider two points $x^1, x^2 \in E^n$. The line segment between them is described by the point

$$w = \theta x^1 + (1 - \theta)x^2$$

as θ varies between 0 and 1. Convex sets have the property that the line segment connecting any two points in the set is also in the set. More precisely, a set $C \subset E^n$ is **convex** if

$$x^1, x^2 \in C$$

implies

$$w = \theta x^1 + (1 - \theta)x^2 \in C \quad \text{for any } \theta \quad 0 \le \theta \le 1$$

For example, E^n is a convex set; other examples are given in Fig. 2.1.

A useful lemma about convex sets states that intersections preserve convexity.

Lemma 2.2: *Let C_i, $i = 1, \ldots, l$, be convex sets. Then the set*

$$C = \bigcap_{i=1}^{l} C_i$$

is also convex.

PROOF: If $x^1, x^2 \in C$, then by definition of intersection $x^1, x^2 \in C_i$, $i = 1, \ldots, l$. As each C_i is convex, for $0 \le \theta \le 1$,

$$w = \theta x^1 + (1 - \theta)x^2 \in C_i \qquad i = 1, \ldots, l$$

Again by intersection

$$w \in C = \bigcap_{i=1}^{l} C_i \quad \blacklozenge$$

Concave and Convex Functions

We can now define concave and convex functions. Given a convex set C, a function h on the set C is **concave** if

$$x^1, x^2 \in C \quad \text{implies}$$

$$h(\theta x^1 + (1 - \theta)x^2) \ge \theta h(x^1) + (1 - \theta)h(x^2) \quad \text{for any } \theta \quad 0 \le \theta \le 1$$

A function h is **convex** if $-h$ is concave.

A useful example of a convex function defined on E^2 is a function that looks like a cereal bowl. An upside down bowl would depict a concave function. A linear function is both concave and convex. Other examples are given in Fig. 2.1, and a special form of the concave function called the strictly concave function is discussed in exercise 2.9.

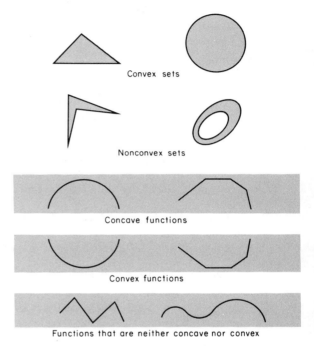

Convex sets

Nonconvex sets

Concave functions

Convex functions

Functions that are neither concave nor convex

Fig. 2.1.

For simplicity, we state the various results of this text in terms of concave functions. The reader should have no difficulty in obtaining the analogous results for convex functions.

One of their most important properties is that the nonnegative sum of concave functions is again concave.

Lemma 2.3: *Let* h_i, $i = 1, \ldots, l$, *each be concave on a convex set C. If* $a_i \geq 0$, $i = 1, \ldots, l$, *the function*

$$h(x) = \sum_{i=1}^{l} a_i h_i(x)$$

is concave on C.

PROOF: By concavity for x^1, $x^2 \in C$, and $w = \theta x^1 + (1 - \theta)x^2$, $0 \le \theta \le 1$

$$h_i(w) \ge \theta h_i(x^1) + (1 - \theta)h_i(x^2) \qquad i = 1, \ldots, l$$

Because the $a_i \ge 0$, the inequality is preserved so that

$$a_i h_i(w) \ge \theta a_i h_i(x^1) + (1 - \theta)a_i h_i(x^2) \qquad i = 1, \ldots, l$$

Summing

$$h(w) = \sum a_i h_i(w) \ge \theta \sum a_i h_i(x^1) + (1 - \theta) \sum a_i h_i(x^2)$$
$$= \theta h(x^1) + (1 - \theta)h(x^2) \quad \blacklozenge$$

Characterizations of Concavity

Should h be differentiable, we have an equivalent definition of concavity.

Lemma 2.4: *Let h be differentiable on E^n. Then h is concave if and only if*

$$h(y) \le h(x) + \nabla h(x)^t(y - x) \quad \text{for any} \quad x \quad \text{and} \quad y.$$

PROOF: Suppose h is concave. Then given $0 \le \theta \le 1$

$$h\left(\theta y + (1 - \theta)x\right) \ge \theta h(y) + (1 - \theta)h(x)$$

For any $\theta > 0$

$$\frac{h\left(\theta y + (1 - \theta)x\right) - h(x)}{\theta} \ge h(y) - h(x)$$

Taking the limit as $\theta \to 0$ but θ remains positive, we get

$$\lim_{\theta \to 0^+} \left(\frac{h\left(\theta y + (1 - \theta)x\right) - h(x)}{\theta} \right) \ge h(y) - h(x)$$

Since h is differentiable and using Equation (2.1), the left side of the above inequality is simply

$$\nabla h(x)^t(y - x)$$

The first part is established.

Now assume

$$h(y) \leq h(x) + \nabla h(x)^t(y - x) \tag{2.4}$$

Then choosing $x = \theta x^1 + (1 - \theta)x^2$ and $y = x^1$ or x^2 respectively, we obtain

$$h(x^1) \leq h\left(\theta x^1 + (1 - \theta)x^2\right)$$
$$+ \nabla h\left(\theta x^1 + (1 - \theta)x^2\right)^t \left[x^1 - \left(\theta x^1 + (1 - \theta)x^2\right)\right] \tag{2.5}$$

and

$$h(x^2) \leq h\left(\theta x^1 + (1 - \theta)x^2\right)$$
$$+ \nabla h\left(\theta x^1 + (1 - \theta)x^2\right)^t \left[x^2 - \left(\theta x^1 + (1 - \theta)x^2\right)\right] \tag{2.6}$$

As

$$x^1 - \left(\theta x^1 + (1 - \theta)x^2\right) = (1 - \theta)(x^1 - x^2)$$

and

$$x^2 - \left(\theta x^1 + (1 - \theta)x^2\right) = -\theta(x^1 - x^2)$$

and letting

$$w = \theta x^1 + (1 - \theta)x^2$$

Equations (2.5) and (2.6) become

$$h(x^1) \leq h(w) + \nabla h(w)^t(1 - \theta)(x^1 - x^2) \tag{2.7}$$

and

$$h(x^2) \leq h(w) - \nabla h(w)^t\theta(x^1 - x^2) \tag{2.8}$$

Multiply (2.7) by $\theta \geq 0$ and (2.8) by $(1 - \theta) \geq 0$, then add to obtain

$$\theta h(x^1) + (1 - \theta)h(x^2) \leq \theta h(w) + (1 - \theta)h(w)$$
$$+ \theta \nabla h(w)^t(1 - \theta)(x^1 - x^2) - (1 - \theta)\nabla h(w)^t\theta(x^1 - x^2) = h(w) \quad \blacklozenge$$

The lemma suggests an alternative definition of concavity for differentiable concave functions. Essentially, this definition states that a linear approximation to a concave function overestimates the function. Observe

that the original definition required a linear interpolation to underestimate
a concave function.

Should h have continuous second partial derivatives, we have a convenient test to verify concavity. As the next lemma shows if h's Hessian
matrix of second partial derivatives is negative semidefinite, then h is concave. Moreover, the converse is also true, and we obtain an equivalent
definition of concavity in terms of the Hessian. Recall that a matrix H is
negative semidefinite if for any x

$$x^t H x \leq 0$$

In the following $H(y)$ indicates the Hessian matrix of second partial derivatives evaluated at y.

Lemma 2.5: *Let h have continuous second partial derivatives. Then h is
concave if and only if the Hessian matrix is negative semidefinite.*

PROOF: The Taylor expansion of h is, for some θ, $0 \leq \theta \leq 1$

$$h(y) = h(x) + \nabla h(x)^t (y - x) + \tfrac{1}{2}(y - x)^t H\left(x + \theta(y - x)\right)(y - x)$$

Clearly,

$$\left(y - x\right)^t H\left(x + \theta(y - x)\right)\left(y - x\right) \leq 0 \qquad (2.9)$$

if and only if

$$h(y) \leq h(x) + \nabla h(x)^t (y - x) \qquad (2.10)$$

If the Hessian is negative semidefinite (2.9) holds. Then from (2.10)
and Lemma 2.4 h is concave.

Now suppose h is concave but that the Hessian is not negative semidefinite for some x. Then for appropriate y

$$(y - x)^t H(x)(y - x) > 0$$

By continuity of the Hessian we may select y so close to x that for all θ,
$0 \leq \theta \leq 1$

$$(y - x)^t H\left(x + \theta(y - x)\right)\left(y - x\right) > 0$$

But then using Equation (2.9), we find Equation (2.10) could not hold.
However Lemma 2.4 requires that (2.10) hold. The contradiction is
evident. ◆

2.3.1. Some Important Implications of Concavity for Nonlinear Programming

Sufficiency for the Unconstrained Problem

Suppose in the unconstrained problem that the objective function is concave, then the necessary conditions for optimality also become sufficient, as the next proposition demonstrates.

Proposition 2.6: *Let h be a differentiable concave function on E^n. Then*

$$\nabla h(x^*) = 0$$

if and only if x^* maximizes h over E^n.

PROOF: Suppose $\nabla h(x^*) = 0$. For any point y, by concavity

$$h(y) \le h(x^*) + \nabla h(x^*)^t(y - x^*) = h(x^*)$$

Thus x^* maximizes h over E^n.
The necessity holds via Corollary 2.1.1.   ◆

Convexity of the Feasible Region with Concave Constraints

Although optimality conditions for the constrained problem have not yet been developed, concavity will play a role in obtaining sufficiency for the constrained case similar to its role in the unconstrained case. The following lemma and proposition will be of great help in this regard. Specifically, if all of the constraints are concave, then the feasible region F will be a convex set.

Lemma 2.7: *Let h be a concave function. Then for any fixed scalar γ the set in E^n*

$$H_\gamma = \{x \mid h(x) \ge \gamma\}$$

is convex.

PROOF: Let $x^1, x^2 \in H_\gamma$. Then

$$h(x^1) \ge \gamma \quad \text{and} \quad h(x^2) \ge \gamma$$

By concavity for $0 \le \theta \le 1$,

$$h\big(\theta x^1 + (1 - \theta)x^2\big) \ge \theta h(x^1) + (1 - \theta)h(x^2) \ge \theta\gamma + (1 - \theta)\gamma = \gamma$$

Hence $\theta x^1 + (1 - \theta)x^2 \in H_\gamma$ and H_γ is convex.   ◆

Proposition 2.8: *Let $g_i \geq 0$, $i = 1, \ldots, m$, be the constraints of a NLP problem. Suppose for each i the set*

$$H_\gamma^i = \{x \mid g_i(x) \geq \gamma\}$$

is convex for any fixed scalar γ. Then the feasible region $F = \{x \mid g_i(x) \geq 0$, $i = 1, \ldots, m\}$ is convex.

PROOF: Set $\gamma = 0$. The feasible set

$$F = \bigcap_{i=1}^{m} H_0^i$$

Since each H_0^i is convex, by Lemma 2.2 F is convex.   ◆

Lemma 2.7 and Proposition 2.8 verify that the feasible region is convex if the constraints are concave functions.

Additional Properties

Several additional properties of concave functions and convex sets are listed (exercise 2.5).

(a) h is concave on E^n if and only if the set

$$\{(x, \gamma) \mid h(x) \geq \gamma\} \subset E^{n+1}$$

is convex, where γ is a scalar and may vary.

(b) A half space $\{x \mid a^t x \geq b\}$ is a convex set for any vector $a \neq 0$ and scalar b.

(c) The linear function $h(x) = a^t x + b$ is both concave and convex.

(d) Let A be an $m \times n$ matrix and b an m vector. Then the set of all x that satisfy

$$Ax = b$$
$$x \geq 0$$

is a convex set.

(e) Let h_i, $i = 1, \ldots, l$, be concave functions. Then the function

$$h(x) = \min \quad [h_1(x), \ldots, h_l(x)]$$

itself is concave.

(f) Let x^i, $i = 1, \ldots, l$, be in a convex set C. Suppose

$$x = \sum_{i=1}^{l} \gamma^i x^i$$

where the γ^i are fixed scalars, $\gamma^i \geq 0$, and

$$\sum_{i=1}^{l} \gamma^i = 1$$

Then $x \in C$.

(g) Suppose h is a concave function. Let

$$x = \sum_{i=1}^{l} \gamma^i x^i$$

where the fixed scalars $\gamma^i \geq 0$ and

$$\sum_{i=1}^{l} \gamma^i = 1$$

Then

$$h(x) \geq \sum_{i=1}^{l} \gamma^i h(x^i)$$

Pseudoconcave and Quasi-Concave Functions

Many applications require some, but not all, of the properties of a concave function. In such situations we may employ functions that preserve the required properties yet are not as restrictive as concave functions.

As mentioned above, two properties of concave functions are especially valuable in NLP. The first is that $\nabla h(x) = 0$ implies x is a global maximum, while the second is that the set

$$\{x \mid h(x) \geq \gamma\}$$

is convex for any scalar γ. The first property ensures that the condition

$$\nabla h(x) = 0$$

is both necessary and sufficient for x to maximize h over E^n. Should each of the constraints possess the second property, then the constraint set is convex by Proposition 2.8. Both properties are quite useful in establishing sufficient conditions for optimality. A function that possesses both of these properties is the pseudoconcave function, while the quasi-concave function possesses only the second property. We consider the pseudoconcave function first.

The Pseudoconcave Function

A differentiable function $h: E^n \rightarrow E^1$ is **pseudoconcave** if

$$\nabla h(x)^t(y - x) \leq 0$$

implies

$$h(y) \leq h(x)$$

Intuitively, whenever a directional derivative indicates a decrease, the function continues to decrease in that direction. A function h is termed **pseudoconvex** if $-h$ is pseudoconcave.

The following lemma, which verifies that a pseudoconcave function has the first property previously mentioned, is left as exercise 2.7.

Lemma 2.9: *Let h be pseudoconcave, then*

$$\nabla h(x) = 0$$

is a necessary and sufficient condition that x maximize h over E^n.

Observe that any concave function is pseudoconcave yet a pseudo-concave function may not be concave (exercise 2.8).

The Quasi-Concave Function

The quasi-concave function will now be defined.

A function $h: E^n \rightarrow E^1$ is called **quasi-concave** if given $x^1, x^2 \in E^n$, for any θ, $0 \leq \theta \leq 1$,

$$h\left(\theta x^1 + (1 - \theta)x^2\right) \geq \min \quad [h(x^1), h(x^2)]$$

We call a function h **quasi-convex** if $-h$ is quasi-concave. The following lemma both establishes that h has the second property previously discussed and provides an equivalent definition for quasi-concavity.

Lemma 2.10: *A function h is quasi-concave if and only if the set*

$$H_\gamma = \{x \mid h(x) \geq \gamma\}$$

is convex for any scalar γ.

PROOF: Let h be quasi-concave. For any given γ select $x^1, x^2 \in H_\gamma$. Then

$$h(x^1) \geq \gamma \quad \text{and} \quad h(x^2) \geq \gamma$$

and by quasi-concavity for $0 \leq \theta \leq 1$

$$h\left(\theta x^1 + (1 - \theta)x^2\right) \geq \min \quad [h(x^1), h(x^2)]$$
$$\geq \gamma$$

Thus $\theta x^1 + (1 - \theta)x^2 \in H_\gamma$ and H_γ is convex.

Suppose now that H_γ is convex for any γ. Given x^1, x^2 assume

$$h(x^1) \geq h(x^2) \tag{2.11}$$

Set

$$\gamma = h(x^2) \tag{2.12}$$

Then by definition

$$x^1, x^2 \in H_\gamma \quad \text{where } \gamma = h(x^2)$$

and because H_γ is convex, for $0 \leq \theta \leq 1$

$$\theta x^1 + (1 - \theta)x^2 \in H_\gamma$$

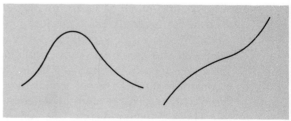

Pseudoconcave functions

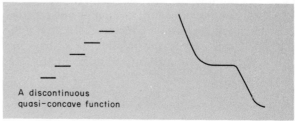

A discontinuous
quasi-concave function

Quasi-concave functions

Fig. 2.2.

But this means

$$h(\theta x^1 + (1 - \theta)x^2) \geq \gamma = \min \quad [h(x^1), h(x^2)]$$

via (2.11) and (2.12).   ◆

Any pseudoconcave function and also any concave function is quasi-concave. Consequently, a pseudoconcave function possesses the second property. However, quasi-concave functions need not be pseudoconcave (see Fig. 2.2 and exercise 2.8). Indeed, if h is quasi-concave and differentiable, the property

$$\nabla h(x) = 0$$

need not imply that x maximizes h over E^n.

Quasi-concave functions and pseudoconcave functions, as well as concave functions, will be extremely useful in the remainder of this book.

2.4. KUHN-TUCKER NECESSARY CONDITIONS

The Kuhn-Tucker (K-T) conditions under reasonable assumptions are necessarily satisfied if x^* is the optimal point to the NLP problem. Recall that Corollary 2.1.1 provides a necessary condition for x^* to be the unconstrained maximum of f. We employ the term unconstrained because x^* maximized f over all of E^n, not just a portion delimited by constraints. The absence of constraints performed a crucial role in the proof of Corollary 2.1.1 as it was possible to move in the direction $d = \nabla f(x^*)$ away from x^*. Were there constraints, moving in the direction $\nabla f(x^*)$ might have violated a constraint, and the proof would then be invalid. Considerable difficulty arises with constraints, and this section is devoted to their consideration.

The K-T conditions are the generalization of Corollary 2.1.1 to constrained problems. Intuitively they state that if one moves away from x^* in any direction, as long as one remains in the feasible region, the objective function cannot increase. To make this notion precise, we begin by first introducing the feasible direction concept. Throughout the discussion all functions are assumed differentiable.

Feasible Directions

Let x be a feasible point. We define a **feasible direction** at x to be any direction d with the property that $x + \tau d$ is in the feasible set F for all τ suffi-

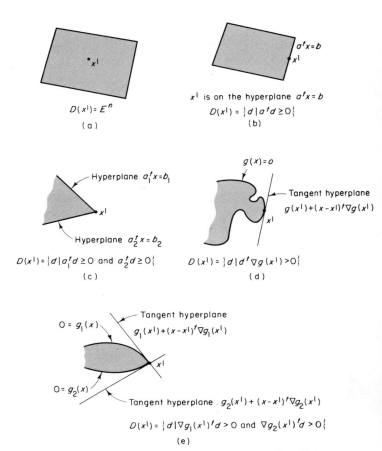

$D(x^1) = E^n$

(a)

x^1 is on the hyperplane $a'x = b$

$D(x^1) = \{d \mid a'd \geq 0\}$

(b)

Hyperplane $a_1'x = b_1$

$D(x^1) = \{d \mid a_1'd \geq 0 \text{ and } a_2'd \geq 0\}$

(c)

$g(x) = 0$

Tangent hyperplane
$g(x^1) + (x - x^1)'\nabla g(x^1)$

$D(x^1) = \{d \mid d'\nabla g(x^1) > 0\}$

(d)

Tangent hyperplane
$g_1(x^1) + (x - x^1)'\nabla g_1(x^1)$

$0 = g_1(x)$

$0 = g_2(x)$

Tangent hyperplane . $g_2(x^1) + (x - x^1)'\nabla g_2(x^1)$

$D(x^1) = \{d \mid \nabla g_1(x^1)'d > 0 \text{ and } \nabla g_2(x^1)'d > 0\}$

(e)

Fig. 2.3.
Examples of feasible direction sets. The feasible set F is shaded lightly.

ciently small. Mathematically, a direction d is feasible at $x = x^1$ if there exists a $\sigma > 0$ such that, any τ, $\sigma \geq \tau \geq 0$, implies $x^1 + \tau d \in F$. The set of all feasible directions at x^1 is denoted by $D(x^1)$.

$$D(x^1) = \{d \mid \exists \ \sigma > 0 \ni \sigma \geq \tau \geq 0 \Rightarrow x^1 + \tau d \in F\}$$

Verbally, a direction d is in $D(x^1)$ if we can move a small distance from x^1 in the direction d and still remain in the feasible set (Fig. 2.3). Here the notation $\exists$ means "there exists," $\ni$ means "such that," while the symbol $\Rightarrow$ means "implies."

Development of Kuhn-Tucker Conditions

The next lemma specifies the behavior of the objective function along directions in $D(x^*)$, where x^* is optimal.

Lemma 2.11: *Let* x^* *be optimal for the NLP problem. Then*

$$\nabla f(x^*)'d \leq 0 \quad \text{for all} \quad d \in D(x^*)$$

PROOF: By contradiction suppose there is a $d^* \in D(x^*)$ such that

$$\nabla f(x^*)'d^* > 0$$

then by Theorem 2.1 a small movement from x^* in the direction d^* would increase x^*. But by taking this movement small enough, since d^* is a feasible direction at x^*, we could generate a feasible point with an increased objective function value. This contradicts the optimality of x^*. ◆

Roughly speaking at an optimal point the directional derivative of the objective function indicates a decrease in any feasible direction. As the next corollary will show this result also holds for all directions in $\bar{D}(x^*)$, the closure of $D(x^*)$.

Recall, a set $\bar{D}$ is the closure of a set D if any point in $\bar{D}$ is the limit of points in D. Although any point in D is also in $\bar{D}$, some points in $\bar{D}$ may not be in D. However, if D is itself a closed set, $D = \bar{D}$. Regrettably $D(x^*)$ need not be closed as certain tangent directions may not be in it (see Fig. 2.3).

Corollary 2.11.1: *Let* x^* *be optimal for the NLP problem. Then*

$$\nabla f(x^*)'d \leq 0 \quad \text{for all} \quad d \in \bar{D}(x^*) \tag{2.13}$$

PROOF: The direction $d^\infty \in \bar{D}(x^*)$ may be expressed as a limit of directions d^k from $D(x^*)$,

$$d^\infty = \lim_{k \to \infty} d^k$$

As $d^k \in D(x^*)$, by Lemma 2.11

$$0 \geq \nabla f(x^*)'d^k \quad \text{for all} \quad k$$

If the limit is taken,

$$0 \geq \lim_{k \to \infty} \nabla f(x^*)'d^k = \nabla f(x^*)'d^\infty$$

and since d^∞ was any point in $\bar{D}(x^*)$ the result holds. ◆

Thus we see that Lemma 2.11 holds for $\bar{D}(x^*)$ as well as $D(x^*)$, and we now are in a position to prove the K-T theorem. The proof of the K-T theorem has two key parts. First, we express the set $\bar{D}(x^*)$ in terms of the constraints. Then via the Farkas lemma, it is possible to reformulate Corollary 2.11.1 using the constraints. The K-T theorem will then follow directly.

Let us first examine $\bar{D}(x^*)$ more closely in our effort to express it explicitly in terms of constraints. At a feasible point x the constraints may be divided into two sets, those that are active at x, $g_i(x) = 0$, and those that are inactive, $g_i(x) > 0$. Let $\mathscr{A}(x)$ be the set of active constraint indices at x

$$g_i(x) = 0 \qquad i \in \mathscr{A}(x)$$
$$g_i(x) > 0 \qquad i \notin \mathscr{A}(x)$$

To formulate $\bar{D}(x)$ in terms of the constraints, only active constraints need be considered. For suppose $g_i(x) > 0$. By continuity of g_i we may move a small distance in any direction from x without violating that constraint. Therefore, the inactive constraints do not influence $\bar{D}(x)$.

The next lemma states that if d is a feasible direction at x, or if $d \in \bar{D}(x)$, then $\nabla g_i(x)^t d \geq 0$ for g_i an active constraint. We first define the set

$$\mathscr{D}(x) = \{d \,|\, \nabla g_i(x)^t d \geq 0 \quad \text{for all} \quad i \in \mathscr{A}(x)\}$$

Lemma 2.12: $\bar{D}(x) \subset \mathscr{D}(x)$.

PROOF: First consider the set of feasible directions $D(x)$. Let $d \in D(x)$, and suppose $g_i(x) = 0$. If $\nabla g_i(x)^t d < 0$, then by Theorem 2.1 for all τ sufficiently small

$$g_i(x + \tau d) < g_i(x) = 0$$

Since such a direction d could not be feasible, $\nabla g_i(x)^t d \geq 0$. Consequently,

$$D(x) \subset \mathscr{D}(x)$$

But the set $\mathscr{D}(x)$ is a closed set, so that (exercise 2.29)

$$\bar{D}(x) \subset \mathscr{D}(x) \quad \blacklozenge$$

Constraint Qualification

Unfortunately situations exist (exercise 2.12) for which there are directions in $\mathscr{D}(x)$ that are not in $\bar{D}(x)$. However, such situations are generally mathematical fabrications and do not seem to arise in practice. There is, thus, little loss in generality by assuming that $\mathscr{D}(x) \subset \bar{D}(x)$ and hence that

$\bar{D}(x) = \mathscr{D}(x)$. This assumption is quite important and is known as the constraint qualification.

If x^* is optimal, then the **constraint qualification** assumes

$$\mathscr{D}(x^*) = \bar{D}(x^*)$$

With the constraint qualification, our first step, that of modifying the set of feasible directions so that it is expressible in terms of constraints, has been achieved. Explicitly

$$\bar{D}(x^*) = \mathscr{D}(x^*) = \{d \mid \nabla g_i(x^*)^t d \geq 0 \quad \text{for all} \quad i \in \mathscr{A}(x^*)\}$$

Summarizing our results so far we obtain from Corollary 2.11.1:

Lemma 2.13: *Let x^* be optimal for the NLP problem. Then under the constraint qualification*

$$\nabla f(x^*)^t d \leq 0 \quad \text{for all} \quad d \in \mathscr{D}(x^*) = \{d \mid \nabla g_i(x^*)^t d \geq 0 \quad \text{for all} \quad i \in \mathscr{A}(x^*)\}$$

Now Farkas' lemma, which is discussed in the Appendix, may be applied. The statement that

$$\nabla f(x^*)^t d \leq 0 \quad \text{for all} \quad d \in \{d \mid \nabla g_i(x^*)^t d \geq 0 \quad \text{for all} \quad i \in \mathscr{A}(x^*)\}$$

is equivalent to the existence of multipliers $\lambda_i \geq 0$, $i \in \mathscr{A}(x^*)$, such that

$$\nabla f(x^*) + \sum_{i \in \mathscr{A}(x^*)} \lambda_i \nabla g_i(x^*) = 0 \tag{2.14}$$

The K-T theorem follows immediately.

Theorem 2.14. *Consider the NLP problem*

$$\max \quad f(x)$$
$$\text{subject to} \quad g_i(x) \geq 0 \qquad i = 1, \ldots, m$$

where all functions are differentiable.

Let x^* be an optimal solution, and assume the constraint qualification holds. Then the following three conditions also hold:

(1)   *x^* is feasible*

There exist multipliers $\lambda_i \geq 0$, $i = 1, \ldots, m$, such that

(2) $\lambda_i g_i(x^*) = 0 \qquad i = 1, \ldots, m$

and

(3) $$\nabla f(x^*) + \sum_{i=1}^{m} \lambda_i \nabla g_i(x^*) = 0$$

PROOF: K-T condition (1) holds trivially, while K-T conditions (2) and (3) are simply reformulations of Equation (2.14). If $i \notin \mathscr{A}(x^*)$, then $g_i(x^*) > 0$, and from K-T condition (2) $\lambda_i = 0$. Consequently, K-T condition (3) is precisely Equation (2.14). ◆

K-T conditions (1), (2), and (3) collectively are called the Kuhn-Tucker (K-T) conditions.

The K-T conditions essentially replace the statement that x^* is optimal by equations and inequalities. From the optimality of x^* it was noted that $\nabla f(x^*)^t d \leq 0$ for all $d \in D(x^*)$ and, in fact, for all $d \in \bar{D}(x^*)$. Then by identifying $\mathscr{D}(x^*)$ with $\bar{D}(x^*)$ the result $\nabla f(x^*)^t d \leq 0$ for all $d \in \mathscr{D}(x^*)$ followed. At this point the Farkas lemma yielded the K-T conditions. Observe, however, that the Farkas lemma actually provides a more powerful result, namely that the K-T conditions are equivalent to the statement

$$\nabla f(x^*)^t d \leq 0 \qquad \text{for all} \qquad d \in \mathscr{D}(x^*)$$

Thus, the K-T conditions are the precise mathematical formulation of the following rough but intuitive concept: In any feasible direction the directional derivative of f must indicate a decrease in the objective function.

A geometrical interpretation of the K-T conditions is given in Figure 2.4. Pictorially, at the optimal point x^* the negative gradient $-\nabla f(x^*)$ lies in the geometric cone of the gradients of active constraints at x^*. By this cone is meant

At the optimal solution x^*, the negative gradient of f is expressible as a nonnegative linear combination of the gradients of the active constraints at x^*. Constraints g_2 and g_3 are active at x^*. Then $-\nabla f(x^*)$ geometrically lies in the cone formed by the gradients of the active constraints at x^*.

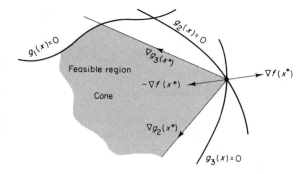

Fig. 2.4.
Geometrical interpretation of the K-T conditions.

$$\left\{ y \,|\, y = \sum_{i \in \mathcal{A}(x^*)} \lambda_i \nabla g_i(x^*),\, \lambda_i \geq 0 \right\}$$

K-T Conditions for Equality Constraints

The NLP problem will often be written with both equality constraints and inequality constraints

$$\max \quad f(x)$$
$$\text{subject to} \quad g_i(x) \geq 0 \qquad i = 1, \ldots, m^1$$
$$\phantom{\text{subject to} \quad} g_i(x) = 0 \qquad i = m^1 + 1, \ldots, m$$

It is left as exercise 2.16 to show that the K-T conditions for this problem are:

(1) $\qquad\qquad\qquad\qquad x^*$ is feasible

There exist multipliers $\lambda_i \geq 0$, $i = 1, \ldots, m^1$, and unconstrained multipliers λ_i, $i = m^1 + 1, \ldots, m$, such that

(2) $\qquad\qquad\qquad \lambda_i g_i(x^*) = 0 \qquad i = 1, \ldots, m^1$

and

(3) $\qquad\qquad\qquad \nabla f(x^*) + \sum_{i=1}^{m} \lambda_i \nabla g_i(x^*) = 0$

Observe that the nonnegative multipliers correspond to the inequality constraints, while the multipliers for the equality constraints are permitted to be negative, positive, or zero. The K-T condition (2) only applies for the inequality constraints.

The K-T conditions are of major importance in NLP and we will utilize them throughout the text.

A Comment

It must be recalled that the K-T conditions can only identify points that are not optimal, as they are necessary and not sufficient conditions. We attack the problem of sufficiency on two fronts. In the next section we will develop sufficient conditions by what is essentially a geometric argument, while later other sufficient conditions will be provided from duality theory.

2.5. SUFFICIENCY AND THE KUHN-TUCKER CONDITIONS

Insufficiency of the K-T Conditions

That the K-T conditions are, by themselves, not sufficient for a point to be optimal is illustrated by the following example. Consider the problem for $x \in E^1$

$$\text{max} \quad x^2$$
$$-1 \leq x \leq 2$$

In terms of the NLP problem $f(x) = x^2$, $g_1(x) = 2 - x$, and $g_2(x) = x + 1$.

The optimal point is obviously $x^* = 2$, yet the K-T conditions hold at $x = -1$, as we now demonstrate.

Certainly K-T condition (1) holds as $x = -1$ is feasible. Now consider K-T conditions (2) and (3) at $x = -1$. These require us to find $\lambda_1 \geq 0$ and $\lambda_2 \geq 0$ such that

$$0 = \lambda_1 g_1(x) = \lambda_1 \cdot 3$$
$$0 = \lambda_2 g_2(x) = \lambda_2 \cdot 0 \tag{2.15}$$

and

$$0 = \nabla f(x) + \lambda_1 \nabla g_1(x) + \lambda_2 \nabla g_2(x) = -2 - \lambda_1 + \lambda_2 \tag{2.16}$$

When these are solved $\lambda_1 = 0$ and $\lambda_2 = 2$. Thus at the point $x = -1$ the K-T conditions hold; yet $x = -1$ is not optimal. Consequently, the K-T conditions may not be sufficient.

A Sufficiency Theorem

We prove that if the objective function f is pseudoconcave and the constraints g_i are quasi-concave, then the K-T conditions are sufficient for optimality.

Theorem 2.15: *In the NLP problem with f and the g_i differentiable, let the objective function f be pseudoconcave and the constraints g_i be quasi-concave. Suppose x^* satisfies the K-T conditions.*

Then x^* is optimal for the NLP problem.

PROOF: Select any feasible point, $y \in F$. It must be proved that

$$f(y) \leq f(x^*)$$

Because the constraints are quasi-concave, the feasible region F is convex. Hence, the line segment between x^* and y is feasible, or if $d^* = y - x^*$, the direction d^* is a feasible direction. By Lemma 2.12 since d^* is feasible,

$$\nabla g_i(x^*)^t d^* \geq 0 \qquad i \in \mathscr{A}(x^*)$$

where $\mathscr{A}(x^*)$ indicates the active constraints at x^*.

Now the K-T conditions are simply the statement that

$$0 \geq \nabla f(x^*)^t d$$

for all d such that $\nabla g_i(x^*)^t d \geq 0$, $i \in \mathscr{A}(x^*)$. But $d = d^*$ satisfies $\nabla g_i(x^*)^t d^* \geq 0$, $i \in \mathscr{A}(x^*)$, and thus

$$0 \geq \nabla f(x^*)^t d^* = \nabla f(x^*)^t (y - x^*)$$

From the pseudoconcavity of f,

$$f(y) \leq f(x^*) \quad \blacklozenge$$

Interpretation

The above theorem possesses an interesting geometric interpretation. The K-T conditions essentially state that in any feasible direction d, the directional derivative of f at x^* indicates a decrease. By pseudoconcavity the objective function decreases in that direction. If we select any feasible point y, because the constraint set is convex, the direction $d^* = y - x^*$ is feasible. Hence, as the directional derivative indicates a decrease in that direction

$$f(y) \leq f(x^*)$$

For a related and more general result see exercise 2.18.

The next section treats duality theorems that are not only important in their own right but can also be used to develop sufficient conditions.

2.6. THE LAGRANGEAN AND DUALITY THEORY

Of direct relevance to the NLP problem is the Lagrangean function

$$L(x, \lambda) = f(x) + \lambda^t g(x) = f(x) + \sum_{i=1}^{m} \lambda_i g_i(x)$$

where

$$\lambda = (\lambda_1, \ldots, \lambda_m)^t, \; g(x) = [g_1(x), \ldots, g_m(x)]^t, \text{ and } \lambda^t g(x) = \sum_{i=1}^{m} \lambda_i g_i(x)$$

Indeed, whenever the Lagrangean has a saddle point the NLP problem is solved. The proof of this result, its implications in duality theory, and the development of additional sufficient conditions are the topics of this section. (For a fundamentally different approach to duality see Chap. 12 in which duality is developed via the penalty algorithm.)

Saddle points

Our development will start by giving a general definition of a saddle point. Any function $K(x, y)$ where $x \in X \subset E^{n_1}$ and $y \in Y \subset E^{n_2}$ is said to have a **saddle point** if there is a point (x^0, y^0) such that

$$K(x, y^0) \leq K(x^0, y^0) \leq K(x^0, y) \quad \text{for all} \quad x \in X \quad \text{and} \quad y \in Y$$

For simplicity it will be assumed in this section that all maxima and minima operations are meaningfully posed.

We now define two important functions that are intimately related to saddle points. The first, a function of y, is obtained by maximizing over x for y fixed. The second instead minimizes over y and is a function of x.

$$K^*(y) = \max_{x \in X} \; K(x, y)$$

and

$$K_*(x) = \min_{y \in Y} \; K(x, y)$$

These functions have many interesting properties. First observe

$$K^*(y) \geq K(x, y) \geq K_*(x) \qquad (2.17)$$

so that

$$K^*(y) \geq K_*(y) \quad \text{for any} \quad (x, y) \qquad (2.18)$$

It will now be shown that the definition of a saddle point may be rephrased in terms of the functions K^* and K_*.

Lemma 2.16. *The following three statements are equivalent*
(a) $\min_{y \in Y} \max_{x \in X} K(x, y) = \max_{x \in X} \min_{y \in Y} K(x, y)$

(b) $K^*(y^0) = K_*(x^0)$ $y^0 \in Y$ $x^0 \in X$

(c) $K(x, y^0) \leq K(x^0, y^0) \leq K(x^0, y)$ $x \in X, y \in Y$

PROOF: Statement (a) may be rewritten

$$\min_{y \in Y} K^*(y) = \max_{x \in X} K_*(x) \tag{2.19}$$

Let y^0 minimize $K^*(y)$ and x^0 maximize $K_*(x)$, then statement (b) holds. By Equation (2.18), (b) implies (2.19) and hence (a).

Through Equation (2.17), (b) is equivalent to

$$K^*(y^0) = K(x^0, y^0) = K_*(x^0) \tag{2.20}$$

and rewriting (2.20) in terms of the definitions, one gets

$$\max_x K(x, y^0) = K(x^0, y^0) = \min_y K(x^0, y)$$

which is simply statement (c). Thus (c) is merely a rephrasing of (b), and the two are equivalent.   ◆

Lemma 2.16 specifies several equivalent ways for identifying a saddle point (x^0, y^0). Regrettably a function may not have a saddle point (exercise 2.20), for it may be that

$$\min_{y \in Y} K^*(y) > \max_{x \in X} K_*(x)$$

With the concepts of the saddle point and the lack thereof defined, we turn to an investigation of the Lagrangean saddle point.

Saddle Points of the Lagrangean

Consider the Lagrangean function for $X \subset E^n$ and

$$\Lambda = \{\lambda = (\lambda_1, \ldots, \lambda_m)^t \,|\, \lambda_i \geq 0, \qquad i = 1, \ldots, m\}$$

The Lagrangean possesses a saddle point if

$$L(x, \lambda^0) \leq L(x^0, \lambda^0) \leq L(x^0, \lambda) \qquad x \in X \qquad \lambda \in \Lambda$$

By analogy to K^* and K_* we define the *primal* function as

$$L_*(x) = \min_{\lambda \in \Lambda} L(x, \lambda)$$

and the *dual* function as

$$L^*(\lambda) = \max_{x \in X} L(x, \lambda)$$

Two problems arise from these functions. The *primal problem* is to determine x^0, called optimal, such that

$$L_*(x^0) = \max_{x \in X} \ L_*(x)$$

Correspondingly, the *dual problem* is to calculate an optimal λ^0 such that

$$L^*(\lambda^0) = \min_{\lambda \in \Lambda} \ L^*(\lambda)$$

The Primal and Dual Problems

The primal problem and the dual problem are basic to the study of NLP, and the remainder of this section is devoted to their analysis. Let us begin study of the primal problem by evaluating the primal function.

$$L_*(x) = \min_{\lambda \in \Lambda} \{f(x) + \sum \lambda_i g_i(x)\} = \begin{cases} f(x) & \text{if all } g_i(x) \geq 0 \\ -\infty & \text{if there exists a } g_i(x) < 0\dagger \end{cases}$$

$$(2.21)$$

because $\lambda_i \geq 0$ for $\lambda \in \Lambda$. Thus if $g_i(x) \geq 0$, setting $\lambda_i = 0$ would minimize. However, if $g_i(x) < 0$, letting $\lambda_i \to +\infty$ minimizes the Lagrangean.

The primal problem requires maximization of the primal function $L_*(x)$. In the maximization process the portion where $L_*(x) = -\infty$ can certainly be disregarded, and the primal problem may be stated as

$$\max_{x \in X} \ L_*(x) = \max \ f(x) \quad \text{over all} \quad x \in X \quad \text{such that}$$
$$g_i(x) \geq 0 \quad i = 1, \dots, m \qquad (2.22)$$

More explicitly the primal problem reduces to

$$\max \ f(x)$$
$$\text{subject to} \quad g_i(x) \geq 0 \quad i = 1, \dots, m \qquad (2.23)$$
$$x \in X$$

Observe a point x is feasible for problem (2.23) if both $g_i(x) \geq 0$, $i = 1, \dots, m$, and $x \in X$. Consequently, problem (2.23) becomes the NLP problem if $X = E^n$. Because of the intimate relationship between the NLP problem and the primal problem, the NLP problem is often referred to as the primal problem. The implicit assumption that $X = E^n$ is, of course, then being made.

---

†From a mathematically rigorous viewpoint the infimum should replace the minimum operation here. The properties of the infimum will be discussed in Chap. 7.

Let us now focus on the dual problem and its relation to the primal. A pair (x^1, λ^1) is said to be feasible for the dual problem if

$$L^*(\lambda^1) = L(x^1, \lambda^1)$$

Should λ^0 be optimal for the dual and (x^0, λ^0) be feasible, then the pair (x^0, λ^0) is also termed optimal.

The next lemma summarizes an important property about the primal and dual problems.

Lemma 2.17: *Suppose x^1 is feasible for problem (2.23). Then*

$$L_*(x^1) = f(x^1)$$

Moreover, if (x^2, λ^2) is feasible for the dual

$$L_*(x^1) = f(x^1) \leq f(x^2) + \sum_{i=1}^{m} \lambda_i^2 g_i(x^2) = L(x^2, \lambda^2) = L^*(\lambda^2) \qquad (2.24)$$

PROOF: If x^1 is feasible for problem (2.23), then via (2.21)

$$L_*(x^1) = f(x^1)$$

Equation (2.18) rewritten in terms of L^* and L_* provides that

$$L_*(x^1) \leq L^*(\lambda^2) \quad \blacklozenge$$

Lemma 2.17 states that the dual problem always provides an upper bound for the primal problem. This holds even if the Lagrangean does not have a saddle point. In such a case

$$f(x^0) < f(x^1) + \sum \lambda_i^1 g_i(x^1)$$

where x^0 is optimal for the primal problem and (x^1, λ^1) is optimal for the dual (exercise 2.21).

Sufficiency

The next theorem provides the key result for determining sufficient conditions for a point to be optimal for the NLP problem.

Theorem 2.18: *Let (x^0, λ^0) be a saddle point to the Lagrangean. That is,*

$$\max_{x \in X} \ L(x, \lambda^0) = L(x^0, \lambda^0) = \min_{\lambda \in \Lambda} \ L(x^0, \lambda)$$

Then x^0 solves the primal problem. Furthermore, if $X = E^n$, then x^0 solves the NLP problem.

PROOF: Reference to Lemma 2.16 reveals that the point x^0 then solves the primal problem. If $X = E^n$ the primal problem reduces to the NLP problem. ◆

The theorem states that if (x^0, λ^0) is a Lagrangean saddle point, x^0 solves the NLP problem where, of course, $X = E^n$. Since it is often difficult to directly verify that a given point (x^0, λ^0) is a saddle point, we now pose several more easily verified conditions that ensure a saddle point exists. Via Theorem 2.18, these conditions themselves become sufficient conditions for a point x^0 to be optimal for the NLP problem.

In the following, $\nabla L(x, \lambda)$ denotes $\nabla f(x) + \sum_{i=1}^{m} \lambda_i \nabla g_i(x)$.

Theorem 2.19: *The point x^1 is optimal for the primal problem if any one of the following conditions holds. Moreover, if $X = E^n$, then any one of these conditions is sufficient for x^1 to be optimal for the NLP problem.*
(a) $L_*(x^1) = L^*(\lambda^2)$ *for some* $x^1 \in X$ *and* $\lambda^2 \in \Lambda$
(b) x^1 *is feasible for problem (2.23), (x^2, λ^2) is feasible for the dual, and*

$$f(x^1) = L(x^2, \lambda^2)$$

(c) 1. x^1 *is feasible for problem (2.23).*

2. $\sum_{i=1}^{m} \lambda_i^1 g_i(x^1) = 0$ *where* $\lambda_i^1 \geq 0$

3. $L(x^1, \lambda^1) = \max_{x \in X} \ L(x, \lambda^1)$
(d) 1. x^1 *satisfies the K-T conditions for the NLP problem, and $x^1 \in X$.*
2. *Letting $\lambda^1 = (\lambda_i^1)$ be the corresponding K-T multipliers*

$$\nabla L(x^1, \lambda^1) = 0 \quad implies \quad L(x^1, \lambda^1) = \max_{x \in X} \ L(x, \lambda^1)$$

(e) *The functions f and g_i, $i = 1, \ldots, m$, are concave, $x^1 \in X$, and x^1 satisfies the K-T conditions.*

PROOF: In all parts of the proof the conclusion follows from Theorem 2.18 after verifying that a saddle point exists.
(a) From Lemma 2.16 this is equivalent to the existence of a saddle point.

(b) Using Lemma 2.17

$$f(x^1) = L_*(x^1) = L^*(\lambda^2)$$

But then we have part (a) above.

(c) From (c)2 and (c)3,

$$f(x^1) = f(x^1) + \sum \lambda_i^1 g_i(x^1) = L(x^1, \lambda^1) = \max_x \; L(x, \lambda^1) = L^*(\lambda^1)$$

Hence, (x^1, λ^1) is dual feasible, and we are reduced to part (b).

(d) Since the K-T conditions hold, x^1 is feasible and $\sum \lambda_i^1 g_i(x^1) = 0$. Moreover, K-T condition (3) is the statement that

$$\nabla L(x^1, \lambda^1) = 0$$

Case (c) is then satisfied.

(e) Because of concavity, Lemma 2.3, and the fact that $\lambda_i^1 \geq 0$, for fixed λ^1 the function

$$L(x, \lambda^1) = f(x) + \sum \lambda_i^1 g_i(x)$$

is a concave function of x. Hence

$$\nabla L(x^1, \lambda^1) = 0 \quad \text{implies}$$
$$L(x^1, \lambda^1) = \max_{x \in E_n} \; L(x, \lambda^1)$$
$$= \max_{x \in X} \; L(x, \lambda^1)$$

since $X \subset E^n$. Then case (d) holds. ◆

Of the five conditions in Theorem 2.19, (d) and (e) are the most useful. This is because many algorithms for solving the NLP problem determine a point x^1 that satisfies the K-T conditions. It is then desired to know if x^1 is optimal. Certainly it need not be. However, if the f and g_i are concave or if

$$\nabla L(x^1, \lambda^1) = 0 \quad \text{implies} \tag{2.25}$$
$$L(x^1, \lambda^1) = \max_{x \in E^n} \; L(x, \lambda^1)$$

then x^1 is indeed optimal for the NLP problem (see exercise 2.28).

For simplicity we frequently state that dual equality exists whenever a saddle point exists, for then the optimal value of the primal equals the optimal value of the dual. Theorem 2.19 provides several conditions that imply the existence of dual equality.

2.6.1. Computational Forms of the Dual

When solving a NLP problem, it is often important to know if dual equality exists. For suppose we are using an algorithm that calculates both primal feasible points x^p and dual feasible points (x^d, λ^d). The algorithm is designed to stop whenever it calculates an x^p and a (x^d, λ^d) such that

$$L(x^d, \lambda^d) - f(x^p) < \epsilon$$

Exploiting Theorem 2.18, we are then very close to optimal. The stopping condition is meaningful if dual equality exists. However, should this not be the case, it is conceivable that

$$\max_{x \in E^n} L_*(x) + \epsilon < \min_{\lambda \in \Lambda} L^*(\lambda)$$

and the stopping condition could never be satisfied.

Because the existence of dual equality is useful in computation, criteria that guarantee its existence, such as those developed in Theorem 2.19, are extremely valuable. Moreover, to facilitate use of these criteria, we now write the dual explicitly for several cases.

The dual problem in general is

$$\min_{\lambda \in \Lambda} L^*(\lambda) \tag{2.26}$$

A more detailed formulation is

$$\min \quad f(x) + \sum_{i=1}^{m} \lambda_i g_i(x)$$

$$\text{subject to} \quad f(x) + \sum_{i=1}^{m} \lambda_i g_i(x) = \max_{x \in X} f(x) + \sum_{i=1}^{m} \lambda_i g_i(x) \tag{2.27}$$

$$\lambda \geq 0$$

The Concave Dual Problem

Suppose $X = E^n$ and all functions are concave. Then

$$\nabla L(x, \lambda) = 0 \quad \text{if and only if} \quad L(x, \lambda) = \max_{x \in E^n} L(x, \lambda)$$

For this situation we obtain the dual to the NLP problem with concave objective function and concave constraints.

$$\min \quad f(x) + \sum_{i=1}^{m} \lambda_i g_i(x)$$

$$\text{subject to} \quad \nabla f(x) + \sum_{i=1}^{m} \lambda_i \nabla g_i(x) = 0$$

$$\lambda_i \geq 0 \qquad i = 1, \ldots, m$$

LP Dual

A special case of the concave dual problem is the linear programming (LP) dual problem. Let the NLP problem be

$$\max \quad q^t x$$
$$\text{subject to} \quad Ax \leq b$$

The concave dual yields

$$\min \quad q^t x + \lambda^t (b - Ax)$$
$$\text{subject to} \quad q^t - \lambda^t A = 0$$
$$\lambda \geq 0$$

However

$$q^t - \lambda^t A = 0$$

implies

$$q^t x - \lambda^t A x = 0$$

The concave dual then reduces to usual dual LP problem

$$\min \quad \lambda^t b$$
$$\text{subject to} \quad A^t \lambda = q$$
$$\lambda \geq 0$$

2.7. COMPARISON OF SUFFICIENCY CONDITIONS

It is quite interesting to compare the two types of conditions for ensuring sufficiency of the K-T conditions. The first, the condition developed in Sec.

2.5, is that f be pseudoconcave and the feasible set be convex. In Sec. 2.6 the second condition,

$$\nabla L(x^*, \lambda^*) = 0 \quad \text{implies} \quad L(x^*, \lambda^*) = \max_x L(x, \lambda^*)$$

was posed.

These two conditions are different. Indeed, it is left as exercise 2.25 to show that the problem for $x \in E^1$

$$\max \quad x^2 + x^3$$
$$x^4 \leq 1$$

satisfies the second condition but not the first.

Conversely, the problem

$$\max \quad e^{-x}$$
$$x \geq 0$$

satisfies the first, although not the second (see exercise 2.26).

Conclusions

This chapter was devoted to developing conditions for testing whether a given point was optimal or not. Essentially the K-T conditions were necessary, and the K-T conditions combined with either assumptions on the form of the functions or a certain Lagrangean condition yielded sufficiency. Unfortunately, there are optimal points that satisfy the K-T conditions yet do not satisfy these sufficiency conditions.

EXERCISES

2.1. Let $x = (1, 1)^t \in E^2$. Define the direction $d = (2, 1)^t$. What is the equation of the line described by

$$y = x + \tau d$$

as τ varies between $-\infty$ and $+\infty$? Consider the same question for $d = (-1, 2)^t$, $d = (1, \frac{1}{2})^t$, $d = (4, 2)^t$. What about $d = (0, 0)^t$?

Now let τ vary between 0 and $+\infty$. Describe the ray generated by $d = (2, 1)^t$ and $d = (-2, -1)^t$ from $x = (1, 1)^t$.

2.2. The $n - 1$ linearly independent hyperplanes

$$\sum_{j=1}^{n} a_{ij}x_j = b_i \qquad i = 1, \ldots, n - 1$$

determine a line in E^n. In the text a line was expressed as

$$y = x + \tau d \quad \text{for} \quad -\infty < \tau < +\infty$$

Calculate x and d so that they determine the same line as the hyperplanes.

2.3. Construct functions such that at a point x^* for which $\nabla f(x^*) = 0$, x^* is a local maximum, but not a global maximum; a saddle point; a global minimum.

2.4. Prove for f, a differentiable function, that if x^* is (a) a local maximum; (b) a global minimum; then $\nabla f(x^*) = 0$.

2.5. Prove the properties of concave functions and convex sets listed in Sec. 2.3.

2.6. Classify the following functions as concave, convex, pseudoconcave, pseudoconvex, quasi-concave, quasi-convex, or none of these.
- (a) e^{-x^2} $x \in E^1$
- (b) e^x $x \in E^1$
- (c) x $x \in E^1$
- (d) x^2 $x \in E^1$
- (e) x^3 $x \in E^1$
- (f) x^{2k} $x \in E^1$ where k is an integer
- (g) x^{2k+1} $x \in E^1$ where k is an integer
- (h) $\sin(x)$
- (i) $\arctan(x)$

2.7. Prove: Let h be pseudoconcave, then

$$\nabla h(x) = 0$$

if x maximizes h over E^n.

2.8. (a) Prove that a concave function is pseudoconcave. Give an example of a pseudoconcave function that is not concave.
- (b) Prove that a pseudoconcave function is quasi-concave but not conversely.
- (c) What does a function h defined on the plane that is both pseudoconcave and pseudoconvex look like?
- (d) Prove that the sum of pseudoconcave functions need not be pseudoconcave.

2.9. A function h is called strictly concave if given $x^1, x^2 \in E^n$, $x^1 \neq x^2$

$$h(\lambda x^1 + (1 - \lambda)x^2) > \lambda h(x^1) + (1 - \lambda)h(x^2)$$

where $0 < \lambda < 1$. Observe the strict inequalities for λ.
  (a) Let x^* maximize h over E^n. Show that x^* is the unique point that maximizes.
  (b) Consider the quadratic form

$$h(x) = q^t x + \tfrac{1}{2} x^t Q x$$

   where Q is a symmetric $n \times n$ matrix.
       Prove that h is strictly concave if and only if Q is negative definite. A matrix Q is negative definite if

$$x^t Q x < 0 \quad \text{for} \quad x \neq 0$$

  (c) Define a strictly convex function.

2.10. If Q is negative semidefinite, show that the functions

$$f_1(x) = x^t Q x$$
$$f_2(x) = q^t x + \tfrac{1}{2} x^t Q x$$

are concave on E^n where q is an n vector. Also prove the converse.

2.11. Consider the constraint set in E^2.

$$x_1 + x_2 \leq 3$$
$$x_1 \qquad \geq 0$$
$$(x_1)^2 + (x_2)^2 \leq 16$$

For each feasible $x = (x_1, x_2)^t$ determine

$$D(x) \qquad \bar{D}(x) \qquad \mathcal{D}(x)$$

2.12. (a) Consider the constraint set

$$(1 - x_1 - x_2)^3 \geq 0$$
$$x_1 \geq 0$$
$$x_2 \geq 0$$

Prove at the point $x^* = (\tfrac{1}{2}, \tfrac{1}{2})^t$ that

$$\bar{D}(x^*) \neq \mathcal{D}(x^*)$$

(b) Show that the constraint set

$$x_1 + x_2 \leq 1$$
$$x_1 \geq 0$$
$$x_2 \geq 0$$

defines precisely the same feasible set as in 2.12(a). Moreover, prove that for any feasible x

$$\bar{D}(x) = \mathcal{D}(x)$$

2.13. Prove that if all constraints are linear, then the constraint qualification holds.

2.14. Let the constraints g_i of the NLP problem be concave and continuously differentiable. Suppose there is a point a such that

$$g_i(a) > 0 \qquad i = 1, \ldots, m$$

Show the constraint qualification holds at any feasible point x.
Hint: Prove $\nabla g_i(x) \neq 0$ $i \in \mathcal{A}(x)$. Take any direction d such that

$$\nabla g_j(x)^t d = 0 \qquad \text{for some} \quad j \in \mathcal{A}(x)$$
$$\nabla g_i(x)^t d \geq 0 \qquad \text{for all} \quad i = 1, \ldots, m$$

Let b be a point on the ray from x in the direction d, and use the points a, x, and b to reduce the problem to E^2 (see Fig. 2.5.).

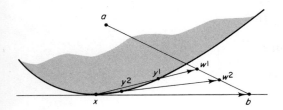

Fig. 2.5.

Construct a sequence of points w^k on the line segment (a, b) converging to b. Let $d^k = w^k - x$. Define y^k as the point in F nearest to w^k on the line segment (x, w^k). Use the convexity of F, and the fact that $\nabla g_i(x) \neq 0, i \in \mathcal{A}(x)$, implies that d is tangent to show that $x \neq y^k$. Hence prove that the entire line segment (y^k, x) is feasible and that the d^k are feasible directions converging to d.

2.15. *The K-T Constraint Qualification.* A continuous function $\varphi \colon E^1 \to E^n$ is said to be an arc in E^n. As the parameter τ varies, $\varphi(\tau)$ describes a path in E^n.

The K-T constraint qualification holds at a point x if, given $d \in \mathscr{D}(x)$, there exists an arc φ with the following three properties.

    (a) there is a $\quad \sigma > 0 \quad$ such that for all τ, $\quad 0 \le \tau \le \sigma$, $\qquad \varphi(\tau) \in F$

    (b) $\varphi(0) = x$

and

    (c) $\displaystyle \lim_{\tau \to 0^+} \frac{\varphi(\tau) - \varphi(0)}{\tau} = d$

In (c) the limit is taken only over positive values of τ.

Prove that if x^* is optimal for the NLP problem with continuously differentiable functions, and if the K-T constraint qualification holds at x^*, then the K-T conditions hold at x^*.

Hint: Since $\varphi(0)$ is optimal, given any $d \in \mathscr{D}(x^*)$ and τ sufficiently small, by Taylor's expansion

$$0 \ge \nabla f[x^* + \theta\{\varphi(\tau) - \varphi(0)\}]^t [\varphi(\tau) - \varphi(0)]$$

for $0 \le \theta \le 1$.

Dividing by $\tau > 0$ and taking limits

$$0 \ge \nabla f(x^*)^t d$$

2.16. Formulate the K-T conditions for the problem

$$\max \quad f(x)$$
$$\text{subject to} \quad g_i(x) \ge 0 \quad i = 1, \ldots, m^1$$
$$g_i(x) = 0 \quad i = m^1 + 1, \ldots, m$$

2.17. *An example of a problem whose optimal point satisfies the K-T conditions yet not the sufficient conditions.*

Consider the problem for $x \in E^1$

$$\max \quad x^4$$
$$\text{subject to} \quad -\tfrac{1}{2} \le x \le 1$$

Show that, although the K-T conditions hold at the optimal point, no sufficiency condition stated in this text is satisfied.

2.18. *Sufficiency of K-T Conditions for Star Shaped Sets.* The feasible set F is said to be *star shaped* at x if $y \in F$ implies $\theta y + (1 - \theta)x \in F$ for $0 \le \theta \le 1$. Suppose the K-T conditions hold at a point x^* and the objective function f is pseudoconcave.

(a) Prove that if F is star shaped at x^*, then x^* is an optimal solution and hence that these conditions are sufficient for optimality.

(b) Demonstrate that F need not be a convex set.

2.19. Let $K(x, y) = xy$ where $x \in E^1$ and $y \in E^1$. Prove that $K(x, y)$ has a saddle point

$$\max_{x \in E^1} \min_{y \in E^1} K(x, y) = \min_{y \in E^1} \max_{x \in E^1} K(x, y)$$

2.20. Construct a $K(x, y)$ where $x \in E^1$ and $y \in E^1$ such that

$$\max_{x \in X} \min_{y \in Y} K(x, y) < \min_{y \in Y} \max_{x \in X} K(x, y)$$

for appropriate $X \subset E^1$ and $Y \subset E^1$.

2.21. Consider the following programming problem for $x \in E^1$

$$\max \quad x^2$$
$$0 \le x \le 1$$

Demonstrate that the corresponding Lagrangean does not have a saddle point.

2.22. Suppose that x^0 is optimal for the NLP problem, that (x^1, λ^1) is optimal for the dual problem, and that

$$f(x^0) = f(x^1) + \sum_{i=1}^{m} \lambda_i^1 g_i(x^1)$$

Show that (x^0, λ^1) is also optimal for the dual problem.

2.23. Suppose x^* is optimal for the NLP problem, the Lagrangean has a saddle point, and λ^* are the corresponding dual multipliers. Prove

$$\lambda_i^* g_i(x^*) = 0 \qquad i = 1, \ldots, m$$

2.24. Consider the NLP problem with quadratic concave objective function and linear constraints.

$$\max \quad q^t x + \tfrac{1}{2} x^t Q x$$
$$\text{subject to} \quad Ax \le b$$

Show that its dual may be written

$$\min \quad \lambda^t b - \tfrac{1}{2} x^t Q x$$
$$\text{subject to} \quad A^t \lambda - Q x = q$$
$$\lambda \ge 0$$

2.25. *A Nonconcave Problem with Dual Equality Holding.* Solve the problem for $x \in E^1$

$$\max \quad x^2 + x^3$$
$$\text{subject to} \quad x^4 \le 1$$

Show that the objective function is not pseudoconcave. Formulate the dual, and show that the optimal value of the primal equals the optimal value of the dual.

2.26. Observe that $x^* = 0$ is the optimal solution to the problem for $x \in E^1$

$$\max \quad e^{-x}$$
$$\text{subject to} \quad x \ge 0$$

Prove that at $x^* = 0$

$$\nabla L(x^*, \lambda^*) = 0 \quad \text{does not imply} \quad L(x^*, \lambda^*) = \max_x \ L(x, \lambda^*)$$

However, show that e^{-x} is pseudoconcave.

2.27. Formulate the K-T conditions and the dual of the problem

$$\min \quad 4(x_1)^2 + 4x_1 x_2 + 2(x_2)^2 + 3x_1 + e^{x_1 + 2x_2}$$

$$\text{subject to} \quad (x_1)^2 + (x_2)^2 \le 10$$

$$- e^{-(x^2)} \le \tfrac{1}{2}$$

$$x_1 \ge 0$$

2.28. Let x^1 satisfy the K-T conditions, and suppose λ^1 is the corresponding multiplier. Assume the Lagrangean $L(x, \lambda^1)$ is pseudoconcave in x, where here λ^1 is fixed. Prove x^1 solves the NLP problem.

2.29. Let B and C be sets in E^n. Suppose C is closed and $B \subset C$. If $\bar{B}$ denotes the closure of B, prove

$$\bar{B} \subset C$$

2.30. *Convex Hulls.* Given a set $B \subset E^n$, we define the convex hull of B to be the set C where

$$C = \left\{ y \,|\, y = \sum_{i=1}^{k} \gamma_i x_i, \quad x_i \in B, \quad \gamma_i \ge 0, \quad \sum_{i=1}^{k} \gamma_i = 1, \quad k = 1, 2, 3, \ldots \right\}$$

Observe k ranges over all positive integers.

(a) Prove C is convex.

(b) If B is convex, then prove $C = B$.

2.31. *Integrating Concave Functions.* Let $h(x, t)$ be concave in x for each t fixed. If the function $p(t)$ is nonnegative, prove that the integral

$$l(x) = \int_E h(x, t)p(t)\, dt$$

is concave. Here E is a set and we assume the integral is well defined.

Observe this result is useful if $p(t)$ is a probability density function.

2.32. *Composition of Concave Functions.* Let $h(x)$ be concave and suppose $p(y)$ is both concave and increasing in y. Prove that the composite function $l(x) = p[h(x)]$ is concave.

If $h(x)$ is concave, prove using composite functions that $\dfrac{1}{h(x)}$ is convex for $h(x) > 0$.

2.33. Write the dual to the concave NLP problem

$$\begin{aligned}
\max \quad & f(x) \\
\text{subject to} \quad & g_i(x) \geq 0 \qquad i = 1, \ldots, m \\
& Ax = b \\
& x \geq 0
\end{aligned}$$

where f and g_i are concave and A is a $m^1 \times n$ matrix.

NOTES AND REFERENCES

§2.1. and 2.2. Further discussions of the gradient, Taylor's expansion, and the topics in these sections may be found in any advanced calculus text, for example APOSTLE; FLEMING. The appendix to this book also contains a brief review of these topics.

§2.3. For detailed treatment of concavity and convexity, see works by EGGLESTON; and by FENCHEL. The quasi-concave function was suggested by NIKAIDO, although its import to NLP was brought out by ARROW and ENTHOVEN. The value of pseudoconcave functions to NLP was emphasized by MANGASARIAN (1965). New extensions of concavity, such as the arcwise concave function, may be found in ZANGWILL (1967b).

§2.4. The presentation in this section is essentially that of ARROW and UZAWA except that we use a simpler although less general constraint qualification.

The original approach of KUHN and TUCKER is presented in exercise 2.15. The discussion of the constraint qualification in exercise 2.12 can be found in ARROW and ENTHOVEN. We present other analyses of the constraint qualification in exercises 2.13 and 2.14. These exercises are quite important, as they reveal simple tests that ensure the constraint qualification holds. Further discussion of exercise 2.13 may be found in ARROW, HURWICZ, and UZAWA (1961); and in ARROW and UZAWA, while exercise 2.14 is based upon SLATER, although the proof is new.

The Farkas Lemma is discussed in the appendix (see also FARKAS).

A somewhat different approach to necessary conditions that actually preceded the K-T conditions is due to F. John, see JOHN; COTTLE (1963a); MANGASARIAN and FROMOVITZ; ABADIE.

For related discussions, see BURGER; FORSYTHE.

§2.5. Theorem 2.15 may be found in MANGASARIAN (1956). However, exercise 2.18 presents a new extension. This extension is an improvement of the sufficiency arguments in ZANGWILL (1967b).

§2.6. and **2.7.** The duality presentation here is interesting because, unlike most previous approaches, it makes no use of convexity. Basically, the presentation is a simplification of some of the introductory arguments in ZANGWILL (1967b). That reference also states theorems based upon minimax arguments for the existence of the saddle point. The approach of this section is related to that of Stoer, although, as mentioned, it does not use convexity as does Stoer, see STOER; MANGASARIAN and PONSTEIN; KARAMARDIAN. Earlier papers also requiring convexity include DORN (1960a), (1960b), (1960c), and (1961); HANSON; HUARD; MANGASARIAN; WOLFE. For a convex approach that exploits the interesting notion of conjugate functions, see KARLIN; ROCKAFELLER. A different approach using convexity is CHARNES, COOPER, and KORTANEK. Symmetric convex duality has been suggested by DANTZIG, EISENBERG, and COTTLE; COTTLE (1963b). An interesting attack without convexity was proposed by RISSANEN, but as yet there seems to be several mathematical flaws in the theorems or proofs of this reference.

ADDITIONAL COMMENTS

An important topic that there is insufficient space to consider is that of minimizing a concave function over a set. This problem is fundamentally different from maximizing such functions and is quite difficult. Applications of this problem to production and inventory problems abound. See, for example, HADLEY (1964); ZANGWILL (1966b), (1966c), (1966d), (1966e) and (1966f). For a more general approach, refer to ZANGWILL (1967c).

3 Applications of the

Kuhn-Tucker Conditions

and Duality Theory

In Chap. 2 we derived both the Kuhn-Tucker (K-T) conditions and duality theory in an effort to determine when a point was optimal. However, the K-T conditions and duality theory are valuable in their own right, and in this chapter we apply them to several interesting and diverse problems. First we interpret duality theory via a competitive market situation. Then the multipliers in the K-T conditions will be discussed as certain marginal changes in the objective function. As another application of duality theory, we will simplify the Geometric Programming (GMP) problem. Finally, we use the K-T conditions to obtain the maximum principle of optimal control.

3.1. AN ECONOMIC INTERPRETATION OF THE LAGRANGEAN AND DUALITY THEORY

The Lagrangean and the existence of a saddle point for it have various economic interpretations. A particularly interesting one results from the competitive interaction between an industry and the market place. The industry attempts to maximize the profit on its product. Opposing this goal is the market, which because it pays the industry, desires to minimize the industry's profit. The market controls the prices paid by the industry for the raw materials the industry buys in the market place.

Let the nonlinear programming problem represent the industry's objective of maximizing its revenue from its product. Thus, the objective function $f(x)$ represents net revenue from selling the industry's product when the industry is operating at a level x. In addition, the industry requires m raw materials but commences production with an amount b_i, $i = 1, \ldots, m$, of raw material i available. Define

$$g_i^1(x)$$

as the quantity of raw material i used when operating the industry at level x. Also let

$$g_i(x) = b_i - g_i^1(x)$$

If

$$g_i(x) > 0$$

then excess raw material remains after production, while should $g_i(x)$ be negative, then insufficient raw material was initially available to achieve production level x. In NLP terms the industry's problem of maximizing the final product revenue becomes

$$\max \quad f(x)$$
$$\text{subject to} \quad g_i(x) \geq 0 \qquad i = 1, \ldots, m$$

The Lagrangean fits easily into this framework. Let $\lambda_i \geq 0$ be the unit price at which raw material i is purchased or sold at the market. Should $g_i(x) > 0$, then the industry can sell excess raw material and realize an additional profit of

$$\lambda_i g_i(x)$$

Similarly if $g_i(x) < 0$, the industry can purchase an amount $g_i(x)$ of that raw material at cost $\lambda_i g_i(x)$. Such a purchase permits it to obtain sufficient raw material to operate at level x. The Lagrangean

$$L(x, \lambda) = f(x) + \sum_{i=1}^{m} \lambda_i g_i(x)$$

represents the total profit to the industry. It includes the final product revenue $f(x)$ plus the net profit from either selling excess raw material or purchasing raw material to overcome deficiencies. Concomitantly, the Lagrangean is the total cost to the market.

For given prices, i.e., λ_i, set by the market, the industry desires to maximize its profits by adjusting its operating level x. Thus the industry forms

$$L^*(\lambda) = \max_{x} \quad L(x, \lambda)$$

which is the dual function.

As the profits are exacted out of the market, the market attempts to adjust prices given the operating level x of the industry. The market then forms the primal function

$$L_*(x) = \min_{\lambda \geq 0} \quad L(x, \lambda)$$

Should dual equality exist, then at such a saddle point there is an economic competitive equilibrium

$$\min_{\lambda \geq 0} \quad L(x^*, \lambda) = \max_{x} \quad L(x, \lambda^*)$$

Observe first, any change in the operating level x by the industry cannot increase profits. Second, no change in prices by the market can decrease profits. Thus the equilibrium situation results in a stable market.

Implications Relative to the K-T Conditions

From the previous theorems it is known that under certain assumptions when dual equality exists the K-T conditions hold. An economic justification of this result can also be given.

The K-T condition (1) states that

$$g_i(x^*) \geq 0$$

Suppose $g_i(x^*) < 0$. The market by indefinitely increasing the price λ_i could then drive the industry's profit to $-\infty$. Consequently, K-T condition (1) must hold at the saddle point.

The K-T condition (2) that

$$\lambda_i g_i(x) = 0$$

also has an economic interpretation. Because the λ_i are prices, $\lambda_i \geq 0$. If $g_i(x) > 0$ and $\lambda_i > 0$, then the market would be able to decrease λ_i slightly, say to $\lambda_i' < \lambda_i$, yielding

$$\lambda_i' g_i(x) < \lambda_i g_i(x)$$

But in this situation the industry's profit would decrease, and any decrease in profit at equilibrium is impossible. Therefore, $\lambda_i g_i(x) = 0$.

Only K-T condition (3) remains. At equilibrium

$$L(x^*, \lambda^*) = \max_x \quad L(x, \lambda^*)$$

and from Corollary 2.1.1 it is known that K-T condition (3) holds because

$$\nabla f(x) + \sum_{i=1}^{m} \lambda_i^* \nabla g_i(x^*) = 0 \tag{3.1}$$

Economically, $\nabla f(x^*)$ is the marginal revenue due to selling the product, and

$$-\sum \lambda_i^* \nabla g_i(x^*)$$

represents the marginal cost of raw materials. At an optimum marginal costs equal marginal revenue. But this is simply the statement of (3.1).

Finally observe that

$$f(x^*) = L(x^*, \lambda^*)$$

Verbally, the net revenue $f(x^*)$ to the industry is identical to its total profit obtained by adding the net revenue $f(x^*)$ to the net profit from the sale or purchase of raw materials. This follows by K-T condition (2). Moreover, at equilibrium if there is excess raw material and

$$g_i(x^*) > 0$$

then $\lambda_i = 0$, and no profit is made on the excess. Note that the industry does not purchase any additional raw material because

$$g_i(x) \geq 0 \quad \text{all} \quad i \tag{3.2}$$

Were it to do so, the market would charge an infinitely large price. The market always adjusts its prices to ensure that (3.2) holds.

3.2. INTERPRETATION OF THE MULTIPLIERS IN THE KUHN-TUCKER CONDITIONS

The λ multipliers for the K-T conditions are frequently termed Lagrange multipliers, dual variables, shadow prices, or imputed costs. They have the interesting property of giving an imputed cost to a constraint.

Consider the problem

$$\max \quad f(x)$$
$$\text{subject to} \quad g_i(x) \leq b_i \quad i = 1, \ldots, m$$

Here the b_i may be viewed as a limitation on the total amount of resource $i, i = 1, \ldots, m$, available.

We are interested in the behavior of the optimal value of the objective function as the right-hand side $b = (b_i)$ is varied. Let $x(b)$ denote the optimal point as a function of the resource availability b. It will be proved that

$$\frac{\partial f(x(b))}{\partial b_i} = \lambda_i \tag{3.3}$$

Therefore, λ_i is the marginal change in the optimal objective function value as b_i is varied. On an intuitive basis, λ_i indicates the approximate increase in the objective function per unit increase in the availability of resource i. The λ_i may also be interpreted as the imputed price of resource i.

Assumptions

For simplicity assume $x(b)$ is a well-behaved function. Specifically, assume $x(b), f$, and g_i are continuously differentiable. Also suppose the constraint qualification is satisfied so that the K-T conditions hold at $x(b)$ as b is varied. Finally, given $b = b'$, assume that if a specific constraint is active at an optimal solution $x(b')$, then that constraint is also active at the solution $x(b)$ for all b sufficiently near b'. The final assumption merely states that if the ith resource is limiting us so that

$$g_i(x(b')) = b'_i$$

then it will still limit us if we vary b slightly.

Proof of Result

To establish Equation (3.3) recall that the K-T conditions for problem (3.2) are

(1) $$g_i(x(b)) \le b_i \qquad i = 1, \ldots, m$$

(2) $$\lambda_i(g_i(x(b)) - b_i) = 0 \qquad i = 1, \ldots, m \tag{3.4}$$

(3) $$\frac{\partial f(x(b))}{\partial x_i} = \sum_{j=1}^{m} \lambda_j \frac{\partial g_j(x(b))}{\partial x_i} \qquad i = 1, \ldots, n \tag{3.5}$$

Using the chain rule, we arrive at

$$\frac{\partial f(x(b))}{\partial b_k} = \sum_{i=1}^{n} \frac{\partial f}{\partial x_i} \frac{\partial x_i}{\partial b_k}$$

From Equation (3.5)

$$\frac{\partial f(x(b))}{\partial b_k} = \sum_{i=1}^{n} \left(\sum_{j=1}^{m} \lambda_j \frac{\partial g_j(x)}{\partial x_i} \right) \frac{\partial x_i}{\partial b_k}$$

$$= \sum_{j=1}^{m} \lambda_j \left(\sum_{i=1}^{n} \frac{\partial g_j(x)}{\partial x_i} \frac{\partial x_i}{\partial b_k} \right) \tag{3.6}$$

We now evaluate Equation (3.6). First suppose constraint j is active at b so that

$$g_j(x(b)) = b_j$$

Then by assumption it will be active in a neighborhood of b. Clearly

$$\frac{\partial g_j}{\partial b_k} = \delta_{jk}$$

where

$$\delta_{jk} = \begin{cases} 0 & \text{if } j \neq k \\ 1 & \text{if } j = k \end{cases}$$

Therefore, if constraint j is active

$$\frac{\partial g_j}{\partial b_k} = \sum_{i=1}^{n} \frac{\partial g_j}{\partial x_i} \frac{\partial x_i}{\partial b_k} = \delta_{jk} \tag{3.7}$$

Now suppose constraint j is inactive so that

$$g_j(x(b)) > b_j \tag{3.8}$$

If b is varied in a sufficiently small neighborhood, constraint j must remain inactive. Consequently, by K-T condition (2)

$$\lambda_j = 0 \tag{3.9}$$

Inserting (3.7) and (3.9) into (3.6)

$$\frac{\partial f(x(b))}{\partial b_k} = \lambda_k$$

which verifies Equation (3.3).

Comment

One practical implication of this result is that we can easily obtain an indication of how the optimal value of the objective function changes as the availability of a resource changes. Specifically, if b_i is increased by one unit, we expect the objective function to increase by about λ_i units. Note that it was not necessary to solve another NLP problem.

3.3. THE DUAL GEOMETRIC PROGRAMMING PROBLEM

An extremely interesting and practical application of duality theory is to GMP. By employing duality theory, we will reduce the rather complicated form of the GMP problem developed in Chap. 1 to an equivalent problem that is much easier to solve.

The Equivalent Problem

In Chap. 1 the following form of the GMP problem was developed.

$$\min \quad \text{Ln } h_0(y)$$
$$\text{subject to} \quad \text{Ln } h_k(y) \leq 0 \qquad k = 1, \ldots, p \qquad (3.10)$$
$$y - Ax = b$$

and

$$h_k(y) = \sum_{i=n_k}^{m_k} e^{y_i} \qquad k = 0, 1, \ldots, p$$

where $x \in E^n$ and $y \in E^{m_p}$ are variables, b is a fixed m_p vector, the $m_p \times n$ matrix A is fixed, and

$$n_0 = 1 \qquad n_1 = m_0 + 1 \qquad n_2 = m_1 + 1 \cdots n_p = m_{p-1} + 1$$

It is the purpose of this section to demonstrate that solving the GMP problem is, under reasonable conditions, equivalent to solving the following problem

$$\max \quad \text{Ln } v(\delta)$$
$$\text{subject to} \quad A^t \delta = 0 \qquad (3.11)$$
$$\delta \geq 0$$

where

$$\lambda_k = \sum_{i=n_k}^{m_k} \delta_i \qquad k = 0, 1, \ldots, p$$

$$\lambda_0 = 1$$

$$v(\delta) = \left[\prod_{i=1}^{m_p} \left(\frac{c_i}{\delta_i} \right)^{\delta_i} \right] \left[\prod_{k=0}^{p} (\lambda_k)^{\lambda_k} \right]$$

and

$$c_i = e^{b_i} \qquad i = 1, 2, \ldots, m_p$$

Here $b = (b_1, \ldots, b_{m_p})^t$ is the vector appearing in problem (3.10), also $\delta = (\delta_1, \delta_2, \ldots, \delta_{m_p})^t$, and $\lambda = (\lambda_0, \lambda_1, \ldots, \lambda_p)^t$. We define $0 \text{ Ln } 0 = 0$ to ensure that $x \text{ Ln } x$ is continuous at $x = 0$. Then $\text{Ln } v(\delta)$ is continuous for $\delta \geq 0$.

Observe that the objective function $\text{Ln } v(\delta)$ is concave (exercise 3.2). Consequently, problem (3.11) is a NLP problem with a concave objective function and linear constraints, and as will be discussed in Chap. 8, is considerably easier to solve than problem (1.6) or (3.10).

Once problem (3.11) is solved, the optimal values of x_i and y_i in problems (1.6) and (3.10) will be shown to be retrievable from knowledge of the optimal δ_i and λ_k via use of the formulae (exercise 3.3)

$$\frac{e^{y_i}}{h_k(y)} = \frac{\delta_i}{\lambda_k} \qquad i = n_k, \ldots, m_k$$

Development of the Dual GMP Problem

We must now demonstrate how to obtain problem (3.11) from problem (3.10). First it must be proved that $\text{Ln}\, h_k(y)$ is a convex function. This result will permit easy application of duality theory.

Recall, since a convex function is the negative of a concave function, a function will be convex if its Hessian matrix of second partial derivatives is positive semidefinite. A matrix $H = (H_{ij})$ is positive semidefinite if, given any vector $s = (s_1, \ldots, s_l)^t$,

$$s^t H s \geq 0$$

or equivalently

$$\sum_{i=1}^{l} \sum_{j=1}^{l} s_i H_{ij} s_j \geq 0$$

Lemma 3.1: *Let* $h(w) = \text{Ln} \sum_{i=1}^{l} e^{w_i}$. *Then* $h(w)$ *is a convex function.*

PROOF: The proof will be established by demonstrating that the Hessian matrix is continuous and positive semidefinite. Clearly, it is continuous. Define

$$h'(w) = \sum_{i=1}^{l} e^{w_i}$$

Then

$$\frac{\partial h(w)}{\partial w_i \partial w_j} = \begin{cases} \left(\displaystyle\sum_{\substack{k=1 \\ k \neq i}}^{l} e^{w_k + w_i} \right) \Big/ h'(w)^2 & \text{if } i = j \\[3mm] -(e^{w_i + w_j}) / h'(w)^2 & \text{if } i \neq j \end{cases}$$

Therefore, for any s_i and s_j

$$h'(w)^2 \sum_{i=1}^{l} \sum_{j=1}^{l} \frac{\partial h(w)}{\partial w_i \partial w_j} s_i s_j$$

$$= \left\{ \sum_{i=1}^{l} \sum_{\substack{k=1 \\ k \neq i}}^{l} (e^{w_k + w_i} s_i^2) - \sum_{i=1}^{l} \left(\sum_{\substack{j=1 \\ i \neq j}}^{l} e^{w_i + w_j} s_i s_j \right) \right\}$$

$$= \sum_{i=1}^{l} \sum_{\substack{j=1 \\ i \neq j}}^{l} e^{w_i + w_j} (s_i^2 - s_i s_j)$$

$$= \sum_{i=1}^{l} \sum_{j=1}^{l} e^{w_i + w_j} (s_i^2 - s_i s_j)$$

where the final equality holds because

$$s_i^2 - s_i s_j = 0 \quad \text{if} \quad i = j$$

Observe that

$$s_i^2 - s_i s_j = (s_i - s_j)^2 - (s_j^2 - s_i s_j)$$

and

$$\sum_{i=1}^{l} \sum_{j=1}^{l} e^{w_i + w_j} s_j^2 = \sum_{i=1}^{l} \sum_{j=1}^{l} e^{w_i + w_j} s_i^2$$

Therefore

$$\sum_{i=1}^{l} \sum_{j=1}^{l} e^{w_i + w_j} (s_i^2 - s_i s_j) = \sum_{i=1}^{l} \sum_{j=1}^{l} e^{w_i + w_j} (s_i - s_j)^2$$

$$- \sum_{i=1}^{l} \sum_{j=1}^{l} e^{w_i + w_j} (s_i^2 - s_i s_j)$$

When rearranged

$$2 \sum_{i=1}^{l} \sum_{j=1}^{l} e^{w_i + w_j} (s_i^2 - s_i s_j) = \sum_{i=1}^{l} \sum_{j=1}^{l} e^{w_i + w_j} (s_i - s_j)^2 \geq 0$$

Consequently,

$$\sum_{i=1}^{I} \sum_{j=1}^{I} \frac{\partial h(w)}{\partial w_i \partial w_j} s_i s_j = \sum_{i=1}^{I} \sum_{j=1}^{I} e^{w_i + w_j} (s_i - s_j)^2 \Big/ 2h'(w)^2 \geq 0 \quad \blacklozenge$$

Applying Lemma 3.1, we observe that the functions

$$\text{Ln } h_k(y) \qquad k = 0, 1, \ldots, p$$

are convex. Adjusting the concave dual problem in Chap. 2 for convex functions, we find that the dual of problem (3.10) becomes

$$\max \quad \text{Ln } h_0(y) + \sum_{k=1}^{p} \lambda_k \text{ Ln } h_k(y) + \delta^t(b + Ax - y)$$

$$\text{subject to} \quad \frac{e^{y_i}}{h_0(y)} - \delta_i = 0 \qquad i = n_0, \ldots, m_0$$

$$\frac{\lambda_k e^{y_i}}{h_k(y)} - \delta_i = 0 \qquad k = 1, \ldots, p \quad \text{and} \quad i = n_k, \ldots, m_k$$

$$A^t \delta = 0 \tag{3.12}$$

$$\lambda \geq 0$$

From Equation (3.12) $x^t A^t \delta = 0$; also when λ_0 is introduced the dual becomes

$$\max \quad \sum_{k=0}^{p} \lambda_k \text{ Ln } h_k(y) + \delta^t(b - y) \tag{3.13}$$

$$\text{subject to} \quad \frac{\lambda_k e^{y_i}}{h_k(y)} - \delta_i = 0 \qquad k = 0, \ldots, p \quad \text{and} \quad i = n_k, \ldots, m_k$$

$$A^t \delta = 0$$

$$\lambda \geq 0$$

$$\lambda_0 = 1$$

We now must modify problem (3.13) to make it identical with problem (3.11). The key difficulty in making this transformation is the fact that $e^y = 0$ is not defined, or equivalently, Ln 0 is not defined. Hence, these situations require special care.

Proposition 3.2: *Let (λ, δ, y) be feasible for problem (3.13). Then*
(a) $\delta \geq 0$

(b) $\sum_{i=n_k}^{m_k} \delta_i = \lambda_k$

and

(c) $\displaystyle\sum_{k=0}^{p} \lambda_k \, \text{Ln} \, h_k(y) + \delta^t(b - y)$

$$= -\sum_{i=1}^{m_p} \delta_i(\text{Ln} \, \delta_i - b_i) + \sum_{k=0}^{p} \lambda_k \, \text{Ln} \, \lambda_k \equiv \text{Ln} \, v(\delta)$$

PROOF: Since (λ, δ, y) is feasible for problem (3.13),

$$\frac{\lambda_k e^{y_i}}{h_k(y)} = \delta_i \tag{3.14}$$

and

$$\lambda_k \geq 0$$

Therefore (a) must hold.

To prove (b), sum (3.14) producing

$$\frac{\lambda_k \displaystyle\sum_{i=n_k}^{m_k} e^{y_i}}{h_k(y)} = \sum_{i=n_k}^{m_k} \delta_i$$

But $h_k(y) = \displaystyle\sum_{i=n_k}^{m_k} e^{y_i}$, so that (b) holds.

Part (c) will be established by demonstrating that

$$\lambda_k \, \text{Ln} \, h_k(y) - \sum_{i=n_k}^{m_k} \delta_i y_i = -\sum_{i=n_k}^{m_k} \delta_i \, \text{Ln} \, \delta_i + \lambda_k \, \text{Ln} \, \lambda_k \tag{3.15}$$

If $\lambda_k = 0$, then by (3.14) $\delta_i = 0$, $i = n_k, \ldots, m_k$, and (3.15) holds. Should $\lambda_k > 0$, then from (3.14) $\delta_i > 0$. Hence via (3.14) again

$$\text{Ln} \, h_k(y) - y_i = -\text{Ln} \, \delta_i + \text{Ln} \, \lambda_k$$

Multiplying by δ_i and summing

$$\left(\sum_{i=n_k}^{m_k} \delta_i\right) \text{Ln} \, h_k(y) - \sum_{i=n_k}^{m_k} \delta_i y_i = -\sum_{i=n_k}^{m_k} \delta_i \, \text{Ln} \, \delta_i + \left(\sum_{i=n_k}^{m_k} \delta_i\right) \text{Ln} \, \lambda_k$$

Then applying part (b) of this theorem, Equation (3.15) and hence (c) holds. ◆

Proposition 3.2 implies that we may write the dual problem (3.13) as

$$\text{max}\quad \text{Ln } v(\delta)$$
$$\text{subject to}\quad A^t\delta = 0$$
$$\delta \geq 0 \tag{3.16}$$
$$\frac{\lambda_k e^{y_i}}{h_k(y)} - \delta_i = 0$$

where

$$\lambda_0 = 1 \quad \text{and} \quad \sum_{i=n_k}^{m_k} \delta_i = \lambda_k$$

Observe that problem (3.16) is identical with problem (3.11) except for the constraints

$$\frac{\lambda_k e^{y_i}}{h_k(y)} - \delta_i = 0$$

It will now be established that, in effect, these constraints are redundant. Recall, because $0 \text{ Ln } 0 = 0$, $\text{Ln } v(\delta)$ is a continuous function for $\delta \geq 0$.

Theorem 3.3: *Assume there is a $\hat{\delta} > 0$ such that*

$$A^t\hat{\delta} = 0 \tag{3.17}$$

Then if δ^* is optimal for problem (3.16) it is also optimal for problem (3.11).

PROOF: Without loss of generality we may assume

$$\sum_{i=1}^{m_0} \hat{\delta}_i = 1 \tag{3.18}$$

Also observe that although δ^* is optimal for problem (3.16), it is also feasible for problem (3.11).

Now let δ be any feasible point for problem (3.11). To prove the theorem we must show

$$\text{Ln } v(\delta^*) \geq \text{Ln } v(\delta)$$

for then δ^* will be optimal for problem (3.11).

Construct a sequence $\{\delta^l\}_{l=1}^{\infty}$ by

$$\delta^l = \left(1 - \frac{1}{l}\right)\delta + \frac{1}{l}\hat{\delta}$$

We first establish that δ^l is feasible for problem (3.16). By choice of δ and $\hat{\delta}$

$$A^t \delta^l = 0 \tag{3.19}$$

$$\sum_{i=1}^{m_0} \delta_i^l = 1 \tag{3.20}$$

and

$$\delta^l > 0 \tag{3.21}$$

Letting

$$\sum_{i=n_k}^{m_k} \delta_i^l = \lambda_k^l \tag{3.22}$$

since $\delta_i^l > 0$ and $\lambda_k^l > 0$, we may define

$$y_i^l = \text{Ln} \frac{\delta_i^l}{\lambda_k^l}$$

Then

$$\frac{\lambda_k^l e^{y_i^l}}{h_k(y^l)} - \delta_i^l = 0 \tag{3.23}$$

because from (3.22)

$$h_k(y^l) = \sum_{i=n_k}^{m_k} e^{y_i^l} = \frac{\sum \delta_i^l}{\lambda_k^l} = 1$$

It follows from (3.19), (3.20), (3.21), (3.22), and (3.23) that δ^l is feasible for problem (3.16).

Because δ^* is optimal for problem (3.16) and δ^l is feasible,

$$\text{Ln } v(\delta^*) \geq \text{Ln } v(\delta^l) \quad \text{for all} \quad l$$

Consequently, taking limits, since $\delta^l \rightarrow \delta$, and by the continuity of $\text{Ln } v(\delta)$

$$\text{Ln } v(\delta^*) \geq \lim_{l \to \infty} \text{Ln } v(\delta^l) = \text{Ln } v(\delta) \quad \blacklozenge$$

Comments

Theorem 3.3 states that some optimal point to problem (3.11) is also optimal for problem (3.16). The existence of $\hat{\delta}$ may be considered as a constraint qualification.

Exploiting Theorem 3.3, we may validate the statements made in the first part of this section. Essentially, under reasonable assumptions duality theory provides that problem (3.10) and its dual problem (3.13) have optimal objective functions that are equal. We may then solve problem (3.11) and thereby determine the optimal value of the objective function for problem (3.13), and hence of the GMP problem.

Theorem 3.3 also provides the following. There exists an optimal (δ^*, λ^*) for problem (3.16), namely the δ^* in Theorem 3.3, that is also optimal for problem (3.11). This (δ^*, λ^*) permits us to determine the optimal x^* and y^* of the GMP problem (1.6) via the formulae

$$\frac{e^{y_i}}{h_k(y^*)} = \frac{\delta_i^*}{\lambda_k^*}$$

The formulae are, of course, constraints of problem (3.16) so that (δ^*, λ^*) and (x^*, y^*) must satisfy them.

In summary, under reasonable conditions, by solving problem (3.11) for (δ^*, λ^*) we obtain the solution to the GMP problem.

3.4. OPTIMAL CONTROL AND THE MAXIMUM PRINCIPLE

As discussed in Chap. 1 the Optimal Control (OC) problem is a dynamic NLP problem over l time periods in which certain control variables have to be selected to maximize an objective function. The K-T conditions for the OC problem have a particularly interesting form and lead to the maximum principle.

The OC problem was previously stated as to

$$\max \quad \sum_{i=1}^{l} f^i(x^i, u^i)$$

$$\text{subject to} \quad g^i(x^i, u^i) + x^i = x^{i+1} \qquad i = 1, \ldots, l$$

$$q(x^{l+1}) = 0$$

where x^1 is given. Here $g^i = (g_1^i, \ldots, g_n^i)^t$, $x^i = (x_1^i, \ldots, x_n^i)^t$, $u^i = (u_1^i, \ldots, u_m^i)^t$, and $q = (q_1, \ldots, q_s)^t$.

Necessary Conditions

If the constraint qualification holds, the K-T necessary conditions that (x, u) solve the OC problem are as follows.

First (x, u) must be feasible. Second, the vectors of unconstrained multipliers $\lambda^i = (\lambda_1^i, \ldots, \lambda_n^i)^t$ and $\sigma = (\sigma_1, \ldots, \sigma_s)^t$ exist that satisfy the following three conditions:

(1) adjoint conditions

$$\nabla_{x^i} f^i(x^i, u^i) + \lambda^i - \lambda^{i-1} + \sum_{j=1}^{n} \lambda_j^i \nabla_{x^i} g_j^i(x^i, u^i) = 0 \qquad i = 2, \ldots, l$$

(2) Transversality conditions

$$\sum_{j=1}^{s} \sigma_j \nabla_{x^{l+1}} q_j(x^{l+1}) - \lambda^l = 0$$

and

(3) Hamiltonian conditions

$$\nabla_{u^i} f^i(x^i, u^i) + \sum_{j=1}^{n} \lambda_j^i \nabla_{u^i} g_j^i(x^i, u^i) = 0 \qquad i = 1, \ldots, l$$

where

$$\nabla_{x^i} f^i = \begin{pmatrix} \dfrac{\partial f^i}{\partial x_1^i} \\ \cdot \\ \cdot \\ \dfrac{\partial f^i}{\partial x_n^i} \end{pmatrix} \quad \text{and} \quad \nabla_{u^i} f^i = \begin{pmatrix} \dfrac{\partial f^i}{\partial u_1^i} \\ \cdot \\ \cdot \\ \dfrac{\partial f^i}{\partial u_m^i} \end{pmatrix}$$

In OC the λ multipliers are often termed adjoint or costate variables.

Hamiltonian

As previously mentioned, the i superscript indicates the time period. The Hamiltonian function H^i at time i is defined as

$$H^i(x^i, u^i, \lambda^i, \lambda^{i-1}) = f^i(x^i, u^i) + (\lambda^i - \lambda^{i-1})^t x^i + (\lambda^i)^t g^i(x^i, u^i)$$

Observe that the adjoint conditions require that the gradient of the Hamiltonian with respect to the state variable x^i be zero, while the Hamiltonian conditions relate to the gradient with respect to the control variable u^i being zero. The Hamiltonian is intimately related to the Lagrangean, as is easily seen (exercise 3.6).

Maximum Principle

Suppose each Hamiltonian function $H^i(x^i, u^i, \lambda^i, \lambda^{i-1})$ were a concave function in u^i for each fixed x^i and λ^i. Then the Hamiltonian condition just stated is equivalent to the following maximum principle.

The Hamiltonian satisfies the maximum principle if for each $i = 1, \ldots, l,$

$$H^i(x^i, u^i, \lambda^i \, \lambda^{i-1}) = \max_u \; H^i(x^i, u, \lambda^i, \lambda^{i-1})$$

where the maximum is over all $u = (u_1, \ldots, u_m)^t$ and $\lambda^0 = 0$.

Under this concavity assumption the necessary conditions for a feasible (x, u) to be optimal are the existence of multipliers λ that satisfy both the adjoint conditions and the maximum principle in addition to the existence of multipliers σ that satisfy the transversality conditions.

EXERCISES

3.1. Consider the problem

$$\max \quad f(x)$$
$$\text{subject to} \quad g_i(x) \leq b_i \qquad i = 1, \ldots, m$$

Suppose constraint j is inactive at x. Then under the assumptions stated in the text

$$\frac{\partial f(x(b))}{\partial b_j} = 0$$

Resource j would economically be termed a free good. Discuss this interpretation. How does the economic discussion of duality theory given in Sec. 3.1 relate to this interpretation.

3.2. Prove that $\mathrm{Ln}\, v(\delta)$ is a concave function for $\delta \geq 0$.

Hint: examine the Hessian matrix of second partial derivatives, and prove it to be negative semidefinite. (Also see p. 121 in DUFFIN, PETERSEN, ZENER.)

3.3 Suppose δ solves problem (3.11). Indicate how to obtain the optimal x_i values of the GMP problem (1.6) from the formulae

$$\frac{\lambda_k e^{y_i}}{h_k(y)} = \delta_i$$

Show that only linear equations need be solved if $\delta_i > 0$ for all i.
Hint: exploit logarithms.

3.4. (DUFFIN, PETERSEN, and ZENER)

$$\max \quad xy^2z^2$$
$$x^3 + y^2 + z \leq 1$$
$$x > 0 \qquad y > 0 \qquad z > 0$$

Hint: Minimize $(xy^2z^2)^{-1}$.

3.5. (DUFFIN, PETERSEN, and ZENER)

$$\min \quad ax + bx^{-2}y^{-3} + cy^4$$
$$x > 0 \qquad y > 0$$

(a) Solve the problem.
(b) What are the relative contributions of each of the three terms to the minimum value of the objective function.

3.6. Indicate how the sum of the Hamiltonian functions H^i is related to the Lagrangean for the OC problem.

NOTES AND REFERENCES

§3.1. The example is new. For other related economic interpretations, see ARROW and HURWICZ (1960); BAUMOL; KOOPMANS; SAMUELSON (1947).

§3.2. This section is similar to that in Chap. 3 of HADLEY (1964); and the preface of ABADIE.

§3.3. The presentation simplifies some notes due to M. Cannon and P. Wolfe. Dr. Cannon indicated how to obtain the dual GMP problem via duality theory instead of the geometric inequality. For the geometric inequality approach, see DUFFIN, PETERSEN, and ZENER.

§3.4. The development of the necessary conditions for the OC problem presented here is not nearly as general as can be found in PONTRYAGIN, *et al.*; HESTENES; and CANON, CULLUM, and POLAK. It has, however, the virtue of being a straightforward application of the K-T conditions.

The papers by CESARI; and by GOLDSTEIN (1965) relate OC and NLP.

4 The First

Convergence Theorem

The previous chapters developed procedures for identifying an optimal point. Essentially, the K-T conditions plus, say, an assumption about concave functions provided a sufficient test for optimality. We must now investigate how to move from a nonoptimal point to an optimal point by use of an algorithm.

In this chapter we study what makes an algorithm converge and prove a key convergence theorem, Convergence Theorem A. As much of this proof rests upon the notion of the point-to-set map, the chapter concludes with a detailed investigation of the closedness and composition properties of these maps.

4.1. THE ALGORITHM

Our focus will be on algorithms or iterative procedures that calculate a sequence of points $\{z^k\}_{k=1}^{\infty}$. At the heart of an algorithm is the recursive process that given a point z^k generates a successor point z^{k+1}. In this manner starting from initial point z^1, we can calculate the entire sequence $\{z^k\}_1^{\infty}$.

The notion of algorithms or iterative procedures just stated is quite general and includes almost any conceivable recursive technique. However, our dominant interest is in algorithms that converge. In other words, the algorithm should either calculate the optimal point or determine the optimal point in a limiting sense.

Finite Convergence

Only in very special cases, such as linear programming (LP) or quadratic programming (QP), can we hope to determine an optimal point in a finite number of computer operations. This is because the Kuhn-Tucker (K-T) conditions for these two problems can be reformulated in terms of a finite number of linear equation systems (exercise 4.1). Nevertheless, due to computer round-off error even these problems may not truly be solvable in a finite number of operations.

For general nonlinear programming (NLP) problems, unless we are lucky and accidentally find an optimal point, convergence only occurs in a limiting sense. It is this general nonlinear case with which we shall be mainly concerned.

4.2 NOTATION

Because algorithmic convergence concerns limits of sequences and subsequences, we need a notation for subsequences. Let $\mathcal{K}$, perhaps written with a superscript, denote an infinite set of positive integers in their natural order. For example,

$$\mathcal{K} = \{1, 3, 7, 10, 13, 15, 19, 25, \ldots\}$$

or

$$\mathcal{K}^1 = \{1, 4, 9, 16, 25, \ldots\}$$

A subsequence of a given sequence $\{z^k\}_1^\infty$ is denoted $\{z^k\}_{\mathcal{K}}$ for an appropriate $\mathcal{K}$. In the above examples

$$\{z^k\}_{\mathcal{K}} = \{z^1, z^3, z^7, z^{10}, z^{13}, z^{15}, z^{19}, z^{25}, \ldots\}$$

and

$$\{z^k\}_{\mathcal{K}^1} = \{z^1, z^4, z^9, z^{16}, z^{25}, \ldots\}$$

The notation $\{z^{k+1}\}_{\mathcal{K}}$ is the subsequence formed by adding 1 to each $k \in \mathcal{K}$. Thus

$$\{z^{k+1}\}_{\mathcal{K}} = \{z^2, z^4, z^8, z^{11}, z^{14}, z^{16}, z^{20}, z^{26} \ldots\}$$

and

$$\{z^{k+1}\}_{\mathcal{K}^1} = \{z^2, z^5, z^{10}, z^{17}, z^{26} \ldots\}$$

We write convergence of subsequences as

$$z^k \longrightarrow z^\infty \qquad k \in \mathcal{K}$$

or

$$\lim_{k \in \mathcal{K}} z^k = z^\infty$$

The limit of a convergent subsequence is frequently indicated by a superscript symbol ∞, while for the limit of the subsequence $\{z^{k+1}\}_{\mathcal{K}}$ we often employ the superscript symbol $\infty + 1$. Explicitly,

$$z^{k+1} \longrightarrow z^{\infty+1} \qquad k \in \mathcal{K}$$

Here the symbol $\infty + 1$ does not mean add one to infinity; it indicates from which subsequence $z^{\infty+1}$ is the limit.

The notation $\mathscr{K}^2 \subset \mathscr{K}$ specifies that $\mathscr{K}^2$ is an infinite subset of $\mathscr{K}$. Thus $\mathscr{K}^2$ could be the subset

$$\{1, 7, 10, 15, 25, \ldots\}$$

and

$$\{z^k\}_{\mathscr{K}^2} = \{z^1, z^7, z^{10}, z^{15}, z^{25}, \ldots\}$$

4.3. THE CONCEPT OF AN ALGORITHM

To illustrate the notion of an algorithm consider the following simple problem.

$$\begin{aligned}
\max \quad & 5 - x \\
\text{subject to} \quad & x \geq 0
\end{aligned} \tag{4.1}$$

where $x \in E^1$. The optimal point is obviously zero.

We now analyze several possible methods for calculating the optimal point. One type of algorithm might use a function $A: E^1 \to E^1$ to generate the successor point z^{k+1} from the point z^k. Thus given z^1, by use of the recursion

$$A(z^k) = z^{k+1}$$

the entire sequence $\{z^k\}_1^\infty$ is generated.

The best possible algorithm for the above problem (4.1) uses the function

$$A(z) = 0 \tag{4.2}$$

The problem is then solved in one step.

Another algorithm of the same type uses the function

$$A(z) = \tfrac{1}{2}z \tag{4.3}$$

Starting at the point 2, the sequence generated would be

$$\{2, 1, \tfrac{1}{2}, \tfrac{1}{4}, \tfrac{1}{8}, \ldots\}$$

For this algorithm, because $z^{k+1} = \tfrac{1}{2}z^k$, at each iteration the next point is

$\frac{1}{2}$ the remaining distance to zero, and the algorithm converges because the limit of the sequence $\{z^k\}$ is the optimal point

$$\lim_{k \to \infty} z^k = 0$$

Point-to-Set Maps

The above class of algorithms that employ a function

$$A: V \longrightarrow V$$

for the recursive process is not sufficiently general, because a function requires that $A(z)$ be a single point in the space V. A more general class of algorithm uses a point-to-set map

$$A: V \longrightarrow V$$

for the recursive process. By **point-to-set map** it is meant that for any point $z \in V$, $A(z)$ is a set in V.

In terms of the algorithm, suppose we are given a point-to-set map $A: V \to V$. Then, given the point z^k, the successor point

$$z^{k+1} \in A(z^k)$$

Furthermore, we may select any $y \in A(z^k)$ as the successor z^{k+1}.

As an example, consider the point-to-set map $A: E^1 \to E^1$, where

$$A(z^k) = \{y \mid z^k - 4 \le y \le z^k + 10\}$$

Here z^{k+1} could be $z^k + 1$, or $z^k - 4$, or any other point in $A(z^k)$. Another example of a map $A: E^1 \to E^1$ is

$$A(z^k) = \{z^k\} \cup \{y \mid z^k + 1 \le y \le z^k + 4\} \cup \{4\}$$

Again z^{k+1} could be any point in $A(z^k)$.

Given a point-to-set map $A: V \to V$, if for each $z \in V$ the map $A(z)$ is a single point in V, then the point-to-set map reduces to a function. In fact, if A is a function, say $A(z^k) = z^{k+1}$, then we may define an equivalent point-to-set map $A(z^k) = \{z^{k+1}\}$. Often in the following discussion a function will be interpreted as if it were the equivalent point-to-set map.

Autonomous versus Nonautonomous Maps

So far the maps A for the algorithms have only depended upon the point z. Such maps are termed autonomous. Frequently maps will be nonautonomous in that they will depend upon the iteration number k as well as the previous points $z^1, \ldots, z^{k-1}$. To explicitly indicate this dependence upon past history we will write

$$A_k \quad \text{or} \quad A_k(z) \quad \text{instead of} \quad A$$

When history may be disregarded and the map depends only upon z, we omit the subscript k and write

$$A \quad \text{or} \quad A(z)$$

An Algorithm

It is now possible to define precisely the concept of an algorithm. An **algorithm** is an iterative process consisting of a sequence of point-to-set maps $A_k: V \longrightarrow V$. Given a point z^1, a sequence of points $\{z^k\}_1^\infty$ is generated recursively by use of the recursion

$$z^{k+1} \in A_k(z^k)$$

where any point in the set $A_k(z^k)$ is a possible successor point z^{k+1}.

The A_k are called the algorithmic maps. Should $A = A_k$ for all k, then A itself is referred to as the algorithmic map. Initially we only consider algorithms that depend upon a single autonomous map A. Algorithms whose maps depend upon past history will be considered later in the text.

Example

Consider again problem (4.1). Let the algorithm be determined by the algorithmic map $A: E^1 \longrightarrow E^1$ where

$$A(z^k) = \{z \mid \tfrac{1}{4}z^k \leq z \leq \tfrac{1}{2}z^k\} \tag{4.4}$$

In words, given a point z^k, the successor point z^{k+1} is any point such that

$$\tfrac{1}{4}z^k \leq z^{k+1} \leq \tfrac{1}{2}z^k$$

The precise sequence $\{z^k\}$, which this algorithm generates, cannot be predicted beforehand. Given $z^1 = 2$, one sequence the algorithm could generate is

$$\{2, \tfrac{7}{8}, \tfrac{3}{8}, \tfrac{1}{8}, \tfrac{1}{16}, \ldots\}$$

Another is

$$\{2, 1, \tfrac{1}{2}, \tfrac{1}{8}, \ldots\}$$

However, any sequence z^k generated by the algorithm must converge to zero. Indeed, from a given initial point z^1, any sequence generated by this algorithm will converge at least as fast as the sequence generated by the algorithm in (4.3).

4.4. CONVERGENCE PROPERTIES

The previous algorithms converged in the sense that a limit point was optimal. To determine what properties an algorithm should have in order to guarantee convergence, let us examine a nonconvergent algorithm for problem (4.1).

One example of a nonconvergent algorithm depends upon a function A where

$$z^{k+1} = A(z^k) = \begin{cases} 1 + \dfrac{(z^k - 1)}{2} & 1 < z^k \\ \tfrac{1}{2}z^k & \text{otherwise.} \end{cases} \qquad (4.5)$$

Given the initial point 5, the algorithm in (4.5) would generate a sequence $\{z^k\}$ as follows

$$\{5, 3, 2, 1\tfrac{1}{2}, 1\tfrac{1}{4}, 1\tfrac{1}{8}, 1\tfrac{1}{16}\}$$

Clearly

$$\lim_{k \to \infty} z^k = 1$$

The sequence $\{z^k\}$ converges to a nonoptimal point. This is no better than if $\{z^k\}$ did not converge at all. In either case we say the algorithm does not converge.

Continuity and Convergence

In order to distill the essence of algorithmic convergence we now examine in detail the algorithms in (4.3) and (4.4) that converge and also the algorithm in (4.5) that does not. First consider the properties these three algorithms share in common. Suppose an initial feasible point z^1 is given to each. All three algorithms then generate a sequence $\{z^k\}$ such that
    (a) the z^k are feasible, and
    (b) $f(z^{k+1}) > f(z^k)$,   if z^k is not optimal.
Thus all algorithms given an initial feasible point generate feasible points, and more important the objective function strictly improves at each iteration. Yet algorithm (4.5) still does not converge.

What, then, is the difference between the convergent and nonconvergent algorithms? Observe that in (4.3) the function

$$A(z) = \tfrac{1}{2}z$$

is continuous. However, the function for algorithm (4.5)

$$A(z) = \begin{cases} 1 + \dfrac{(z-1)}{2} & 1 < z \\ \tfrac{1}{2}z & \text{otherwise,} \end{cases}$$

is discontinuous at the point $z = 1$ (see Fig. 4.1). It is precisely this discontinuity that causes the difficulty. Indeed, continuity or an analogous property will be later shown to be of great importance in proving convergence.

Algorithm (4.3)

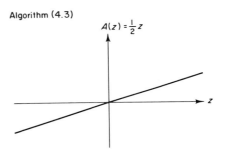

Algorithm (4.5)

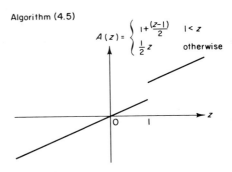

Fig. 4.1.

A Closed Map

Algorithm (4.4), which is a point-to-set map, also converges. As might be expected the point-to-set map for algorithm (4.4) possesses a property

called closedness, which is an extension of the function-continuity concept
to maps. To introduce the concept of closed maps let us review the definition
of continuity. Usually the definition of a function A continuous at z^∞ appears
as

$$\lim_{k \in \mathscr{K}} z^k = z^\infty \quad \text{implies} \quad \lim_{k \in \mathscr{K}} A(z^k) = A(z^\infty)$$

In order to better illustrate the analogy between continuity and closedness,
continuity is rephrased as follows:

    If

    (a) $z^k \longrightarrow z^\infty \qquad k \in \mathscr{K}$

and

    (b) $y^k = A(z^k) \qquad k \in \mathscr{K}$

then

    (c) $y^k \longrightarrow y^\infty \qquad k \in \mathscr{K}$

and

    (d) $y^\infty = A(z^\infty)$

    The definition of a closed map may now be posed. A point-to-set map
$A: V \longrightarrow V$ is **closed** at z^∞ if

    (a) $z^k \longrightarrow z^\infty \qquad k \in \mathscr{K}$

    (b) $y^k \in A(z^k) \qquad k \in \mathscr{K}$

and

    (c) $y^k \longrightarrow y^\infty \qquad k \in \mathscr{K}$

imply

    (d) $y^\infty \in A(z^\infty)$

The map is said to be closed on $X \subset V$ if it is closed at each $z \in X$, and we
say a map is closed if it is closed at each point where it is defined. An equivalent definition for a closed map is considered in exercise 4.2.

    Figure 4.2 is a graphical description of a closed map, and using the
figure, we may interpret the essential notion of closedness. Suppose

$$z^k \longrightarrow z^\infty$$

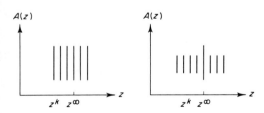

Fig. 4.2.
Closed maps. The set $A(z)$ is depicted as a line segment.

Select $y^k \in A(z^k)$ such that

$$y^k \longrightarrow y^\infty$$

Then the limit y^∞ is in $A(z^\infty)$. Roughly speaking, the set $A(z^\infty)$ must be at least as large as its neighboring sets $A(z^k)$. This must be true for $A(z^\infty)$ to "catch" any limit point y^∞.

A map that is not closed appears in Fig. 4.3. Let $z^k \longrightarrow z^\infty$. Choose $y^k = 1\frac{3}{4} \in A(z^k)$, for all k, then $y^k \longrightarrow y^\infty = 1\frac{3}{4}$. But $1\frac{3}{4}$ is not in $A(z^\infty)$. Hence the A of Fig. 4.3 is not closed at z^∞.

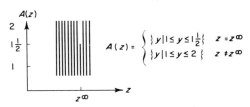

Fig. 4.3.
A map that is not closed.

4.5. A CONVERGENCE THEOREM

Several of the key notions that guarantee convergence have been developed and will now be formulated in a theorem. Before stating this theorem, however, we will prove a lemma.

Lemma 4.1: *Let $Z: X \to E^1$ be a continuous function. Suppose there is a sequence $\{z^k\}_1^\infty$ in X such that*
    (a) $Z(z^k) \leq Z(z^{k+1})$ *all* k
and
    (b) $z^k \longrightarrow z^\infty$ $k \in \mathcal{K}$
Then

$$\lim_{k \to \infty} Z(z^k) = \lim_{k \in \mathcal{K}} Z(z^k) = Z(z^\infty)$$

PROOF: Continuity of Z provides

$$\lim_{k \in \mathcal{K}} Z(z^k) = Z(z^\infty) \tag{4.6}$$

From (a) the sequence $\{Z(z^k)\}_1^\infty$ is monotonic so that for any l

$$Z(z^l) \leq Z(z^\infty) \tag{4.7}$$

Using (4.6) and the definition of limit given $\epsilon > 0$, there is a $k_\epsilon \in \mathcal{K}$ such that for $k \geq k_\epsilon$ and $k \in \mathcal{K}$

$$Z(z^\infty) - Z(z^k) < \epsilon \tag{4.8}$$

For $l \geq k_\epsilon$ by (a)

$$Z(z^l) \geq Z(z^{k_\epsilon}) \tag{4.9}$$

Equations (4.7), (4.8), and (4.9) then yield

$$|Z(z^\infty) - Z(z^l)| < \epsilon$$

for all $l \geq k_\epsilon$. ◆

The lemma may be roughly interpreted as follows. If a sequence is monotonically increasing, that is $Z(z^k) \leq Z(z^{k+1})$, and a subsequence of it converges to some limit point, then the entire sequence converges to that limit.

Compactness

One other notion must be introduced, that of compactness. A set X is said to be compact if any sequence (or subsequence) contains a convergent subsequence whose limit is in X. More explicitly, given a subsequence $\{z^k\}_{\mathcal{K}}$ in X, where X is **compact**, there exists a $\mathcal{K}^1 \subset \mathcal{K}$ such that

$$z^k \longrightarrow z^\infty \qquad k \in \mathcal{K}^1$$

and z^∞ is in X.

In Euclidean spaces it can be shown that compact sets correspond to closed and bounded sets. Thus a compact set must contain all of its edges and cannot be extended to infinity in any direction. The points generated by most algorithms can be contained in such sets.

The Convergence Theorem

The convergence theorem will be stated in quite general terms. Instead of the objective function, we employ a more general Z function. Often Z is the objective function, although other examples will be given in later chapters. Furthermore, in place of optimal points we substitute the broader concept of solution point. A **solution point** is any point in a given set Ω

called the solution set. The algorithm, instead of seeking an optimal point, will seek a solution point.

Although it need not be, the solution set Ω could be defined to be the set of all optimal points, and in that case the algorithm would seek an optimal point. Usually, however, the definition of solution set will depend upon both the problem and the algorithm. For example, in certain problems optimal points may not be of interest. Instead our goal might be to determine a feasible point that gives the objective function f a value at least m. In this situation the solution set is

$$\Omega = \{x \,|\, f(x) \geq m, \, x \text{ feasible}\}$$

Other examples include defining Ω to be the roots of a system of equations, saddle points, economic-equilibrium points, stability points of difference equations, etc. Thus we see that the concept of algorithm developed here, although it is posed in a NLP framework, is considerably broader. However, no matter how Ω is defined it is assumed that a set Ω is given. The algorithm then attempts to determine a point in Ω.

In the theorem the algorithm given a point z^1 generates the sequence $\{z^k\}_1^\infty$ by use of the recursion $z^{k+1} \in A(z^k)$.

Convergence Theorem A

Let the point-to-set map $A: V \longrightarrow V$ *determine an algorithm that given a point* $z^1 \in V$ *generates the sequence* $\{z^k\}_1^\infty$. *Also let a solution set* $\Omega \subset V$ *be given.*

Suppose
(1) *All points* z^k *are in a compact set* $X \subset V$.
(2) *There is a continuous function* $Z: V \longrightarrow E^1$ *such that:*
   (a) *if* z *is not a solution, then for any* $y \in A(z)$

$$Z(y) > Z(z)$$

   (b) *If* z *is a solution, then either the algorithm terminates or for any* $y \in A(z)$

$$Z(y) \geq Z(z)$$

and
(3) *The map* A *is closed at* z *if* z *is not a solution.*
   Then either the algorithm stops at a solution, or the limit of any convergent subsequence is a solution.

PROOF: If the algorithm stops, then by 2(b) it terminates at a solution. Hence, suppose an infinite sequence is generated.

Observe first that by 1 there must be a convergent subsequence

$$z^k \longrightarrow z^\infty \qquad k \in \mathscr{K}$$

Using 2, we see that

$$Z(z^{k+1}) \geq Z(z^k)$$

But then Lemma 4.1 implies

$$\lim_{k \to \infty} Z(z^k) = Z(z^\infty) \qquad\qquad (4.10)$$

Consider now the subsequence $\{z^{k+1}\}_{\mathscr{K}}$. By 1 there is a $\mathscr{K}^1 \subset \mathscr{K}$ such that

$$z^{k+1} \longrightarrow z^{\infty+1} \qquad k \in \mathscr{K}^1$$

and again from Lemma 4.1

$$\lim_{k \to \infty} Z(z^{k+1}) = Z(z^{\infty+1}) \qquad\qquad (4.11)$$

Equations (4.10) and (4.11) then yield

$$Z(z^{\infty+1}) = Z(z^\infty) \qquad\qquad (4.12)$$

To complete the proof let us assume z^∞ is not a solution and establish a contradiction. Certainly
   (a) $z^k \longrightarrow z^\infty$ $\qquad k \in \mathscr{K}^1$
   (b) $z^{k+1} \in A(z^k)$ $\qquad k \in \mathscr{K}^1$
and
   (c) $z^{k+1} \longrightarrow z^{\infty+1}$ $\qquad k \in \mathscr{K}^1$
Then by 3

$$z^{\infty+1} \in A(z^\infty) \qquad\qquad (4.13)$$

Since z^∞ is assumed not a solution, 2(a) ensures that

$$Z(z^{\infty+1}) > Z(z^\infty) \qquad\qquad (4.14)$$

However, since (4.12) and (4.14) are contradictory, z^∞ must be a solution. ◆

Discussion of Convergence Theorem A

The first convergence theorem has now been stated. Although other more powerful convergence theorems will be developed later, Convergence

Theorem A is quite broad in its coverage and we will use it in subsequent chapters to prove the convergence of numerous algorithms.

Let us now examine the assumptions of Convergence Theorem A in more detail. First observe that we utilized the variable z instead of the variable x. The variable z denotes the point on which the algorithmic map depends. Frequently z is actually the variable of the programming problem x, and in that case $x = z$. More generally, however, the variable of the problem and z are different. In fact, as will be demonstrated, one of the key aspects of proving convergence is the determination of the appropriate variable z on which the algorithmic map depends.

Condition 1 of the theorem guarantees that all subsequences are well behaved. Without condition 1 a subsequence might have a limit outside the set V, or perhaps the subsequence might diverge to $+\infty$. Condition 1 prohibits these anomalies. Note, however, that the feasible set F need not be compact, just that the points generated must be on a compact set.

The use of Z in condition 2 is, as mentioned previously, more general than the use of the objective function. One essential ingredient of convergence is that the procedure must adapt or improve. Condition 2(a) requires improvement at each iteration of the algorithm until a solution point is reached. We may interpret the Z function as an adaption function that indicates the algorithm's progress. Many algorithms use the objective function f to measure their progress, and in that case $f = Z$. Some algorithms, however, solve the NLP problem by transforming it into another problem, and the function Z can then be considered as the objective function for the transformed problem. Generally, as Z increases, the algorithm is approaching the solution.

As noted in the examples something similar to condition 3 is required to prohibit the discontinuities that may cause nonconvergence. Observe, however, that we do not require the closedness of the map at a solution point. At a solution point condition 2(b) dominates, and the behavior of the map A is thus immaterial.

In establishing convergence, the proof of condition 3 is usually the most challenging. Often the following heuristic aids in determining a mathematical verification of 3. Let $y \in A(z)$ so that y is a successor to z. Now, perturb z slightly to z', and let $y' \in A(z')$. If y' is close to y, then A probably has the continuity property (i.e., closedness). Basically, a slight change in z should produce a slight change in z's successor. If this statement seems correct on an intuitive level, condition 3 can usually be mathematically verified.

Convergence Theorem A thereby specifies a precise procedure for proving convergence. First the map A and points z must be determined. The compactness property, condition 1, is generally assumed to hold, as it usually does in practice. Then an adaption function Z must be specified and condition 2 proved. Finally, A must be shown to be closed at any point that is not a solution.

4.6. DECOMPOSITION OF THE ALGORITHMIC MAP

Proving directly that an algorithmic map A is closed is often difficult because A may be composed of several parts. In this section we will show how to simplify the proof of A's closedness by instead verifying that each part of A is a closed map. For example, in Chap. 5 the map A will be decomposed into two maps D and M. The D map, given z^k, will determine a direction, while the M map calculates z^{k+1} using the given direction, and the map A will be written

$$A = MD$$

to indicate that A is composed of two maps. Proving that A is closed will only require proving that both M and D are closed. Generally, this latter task is straightforward, and considerably easier than attempting to prove A's closedness directly.

Map Composition

To make these concepts precise, we must develop the notion of map composition. First we recall the definition of a composite function. Suppose two functions $c\colon W \longrightarrow X$ and $b\colon X \longrightarrow Y$ are given, then the composition $a = bc\colon W \longrightarrow Y$ is defined as

$$a(w) = b[c(w)]$$

We will express the composition of two maps analogously. Let $C\colon W \longrightarrow X$ and $B\colon X \longrightarrow Y$ be point-to-set maps. For each $x \in C(w)$, $B(x)$ is a set in Y. The composition, $A(w) = B[C(w)]$, is all the sets $B(x)$ obtained from all the x in $C(w)$.

More precisely the **composite** map $A = BC\colon W \longrightarrow Y$ is defined as

$$A(w) = \cup \{B(x) \,|\, x \in C(w)\}$$

As an example define $B\colon E^1 \longrightarrow E^1$ and $C\colon E^1 \longrightarrow E^1$ by

$$B(x) = \{y \,|\, x - 1 \le y \le x + 4\}$$

and

$$C(w) = \{x \mid w \leq x \leq 1\}$$

Then $A: E^1 \rightarrow E^1$ is

$$A(w) = B[C(w)] = \begin{cases} \{y \mid w - 1 \leq y \leq 5\} & \text{if} \quad w \leq 1 \\ \phi & \text{if} \quad w > 1 \end{cases}$$

Here ϕ denotes the null set. Observe $A(w) = \phi$ if $w > 1$ because in that case $C(w) = \phi$.

Composition Preserves Closedness

Suppose an algorithmic map A is expressible as the composition of several maps each of which is closed. The next lemma verifies that under reasonable conditions A is also closed. In addition, the lemma may be considered as a generalization of the theorem that the composition of two continuous functions is itself continuous. Briefly, the lemma proves that composition preserves closedness.

Lemma 4.2: *Let $C: W \rightarrow X$ and $B: X \rightarrow Y$ be point-to-set maps. Suppose C is closed at w^∞, and B is closed on $C(w^\infty)$. Also assume if $w^k \rightarrow w^\infty$, $k \in \mathcal{K}$, and if $x^k \in C(w^k)$, $k \in \mathcal{K}$, that for some $\mathcal{K}^1 \subset \mathcal{K}$*

$$x^k \longrightarrow x^\infty \qquad k \in \mathcal{K}^1$$

Then the composition $A = BC$ is closed at w^∞.

PROOF: Let $w^k \rightarrow w^\infty$, $y^k \rightarrow y^\infty$, and $y^k \in A(w^k)$ $k \in \mathcal{K}$. We must prove $y^\infty \in A(w^\infty)$.

Select $x^k \in C(w^k)$ such that

$$y^k \in B(x^k) \qquad k \in \mathcal{K}$$

By hypothesis

$$x^k \longrightarrow x^\infty \qquad k \in \mathcal{K}^1$$

As C is closed

$$x^\infty \in C(w^\infty)$$

Similarly

$$y^\infty \in B(x^\infty)$$

Thus

$$y^\infty \in B[C(w^\infty)] = A(w^\infty) \quad \blacklozenge$$

Regrettably Lemma 4.2 does not quite state that the composition of two closed maps is closed. An additional assumption is required to ensure that some subsequence of the $\{x^k\}$ sequence converges. The reader can easily verify in exercise 4.8 that this assumption is mandatory. Other forms of this assumption are posed in the next two corollaries. Corollary 4.2.1 is especially useful. The proof of the corollaries is left as exercise 4.10.

Corollary 4.2.1: *Let* $C: W \longrightarrow X$ *and* $B: X \longrightarrow Y$ *be point-to-set maps. Suppose* C *is closed at* w^∞ *and* B *is closed on* $C(w^\infty)$. *If* X *is compact, then* $A = BC: W \longrightarrow Y$ *is closed at* w^∞.

Corollary 4.2.2: *Let* $C: W \longrightarrow X$ *be a function and* $B: X \longrightarrow Y$ *be a point-to-set map. Assume* C *is continuous at* w^∞ *and* B *is closed at* $C(w^\infty)$. *Then the point-to-set map* $A = BC: W \longrightarrow Y$ *is closed at* w^∞.

Closedness of Arithmetic Compositions

Lemma 4.2 and its corollaries may be applied to several important maps. Let $B: W \longrightarrow E^n$ and $C: W \longrightarrow E^n$ be point-to-set maps.

The sum map $A = B + C: W \longrightarrow E^n$ is defined by

$$A(w) = \{b + c \mid b \in B(w), c \in C(w)\}$$

For example, if $B(w) = \{x \mid 0 \leq x \leq 1\} \subset E^1$ and $C(w) = \{x \mid 1 \leq x \leq 2\} \subset E^1$ then

$$A(w) = \{x \mid 1 \leq x \leq 3\}$$

The inner-product map $A = B^t C: W \longrightarrow E^1$ is defined by

$$A(w) = \{b^t c \mid b \in B(w) \quad \text{and} \quad c \in C(w)\}$$

The set A is the set of all vector inner products $b^t c$ for all $b \in B(w)$ and all $c \in C(w)$.

If $B: W \longrightarrow E^1$, then another form of the product map is the scalar-vector product map

$$A = B \cdot C: W \longrightarrow E^n$$

where

$$A(w) = \{bc \mid b \in B(w), c \in C(w)\}$$

Here, of course, b is a scalar, c is a vector, and bc is a vector.

As might be expected, these three operations yield closed maps but, as in Lemma 4.2, only under an additional assumption that ensures appropriate behavior of the sequences.

Each of the three maps can be expressed as the composition of two maps. For example consider the sum map. First define the point-to-set map $S: W \longrightarrow E^{2n}$ by

$$S(w) = \{(b, c) \mid b \in B(w), c \in C(w)\}$$

Here $b \in E^n$, $c \in E^n$, and $(b, c) \in E^{2n}$. If both B and C are closed maps, then clearly S is closed also. The sum map may then be expressed in terms of map composition by defining the continuous function $h: E^{2n} \longrightarrow E^n$.

$$h(b, c) = b + c$$

Then the sum map $A(w)$ becomes the composition

$$A = hS$$

The next lemma follows immediately from Corollary 4.2.1.

Lemma 4.3: *Let $B: W \longrightarrow Y \subset E^n$ and $C: W \longrightarrow Y \subset E^n$ be closed maps at w^∞. Suppose Y is compact. Then the sum map is closed at w^∞.*

Certainly, analogous lemmas hold for the two vector maps. However, because these three maps have a special structure, weaker hypotheses than compactness can ensure appropriate behavior of sequences. The reader is asked to prove the following lemmas in exercise 4.11.

Lemma 4.4: *Let $B: W \longrightarrow Y \subset E^n$ and $C: W \longrightarrow Y \subset E^n$ be closed maps at w^∞. Then the sum map $A = B + C: W \longrightarrow Y \subset E^n$ is closed at w^∞ if any one of the following three conditions holds:*
  (a) *Either B or C is a continuous function at w^∞.*
  (b) *If $w^k \longrightarrow w^\infty$ and $b^k \in B(w^k)$, $k \in \mathcal{K}$, then there exists a $\mathcal{K}^1 \subset \mathcal{K}$ such that*

$$b^k \longrightarrow b^\infty \qquad k \in \mathcal{K}^1$$

  (c) *Y is compact.*

Lemma 4.5: *Let $B: W \longrightarrow Y \subset E^n$ and $C: W \longrightarrow Y \subset E^n$ be closed maps at w^∞. Then the inner-product map $A = B^t C: W \longrightarrow E^1$ is closed at w^∞ if any one of the following three properties holds:*
  (a) *Both B and C are continuous functions at w^∞.*
  (b) *If $w^k \longrightarrow w^\infty$, $b^k \in B(w^k)$, and $c^k \in C(w^k)$, $k \in \mathcal{K}$, then there exists a $\mathcal{K}^1 \in \mathcal{K}$ such that*

$$b^k \longrightarrow b^\infty \quad and \quad c^k \longrightarrow c^\infty \quad k \in \mathcal{K}^1$$

(c) *Y is compact.*

Lemma 4.6: *Let* $B: W \to V \subset E^1$ *and* $C: W \to Y \subset E^n$ *be closed maps at* w^∞. *Then the scalar-vector product map* $A = B \cdot C: W \to E^n$ *is closed at* w^∞ *if any one of the following five properties holds:*

(a) *Both B and C are continuous functions at* w^∞.

(b) *If* $w^k \to w^\infty$ *and* $c^k \in C(w^k)$, $k \in \mathcal{K}$, *then there exists a* $\mathcal{K}^1 \subset \mathcal{K}$ *such that*

$$c^k \longrightarrow c^\infty \quad k \in \mathcal{K}^1$$

where at least one component of the vector c^∞ *is not zero.*

(c) *If* $w^k \to w^\infty$ *and* $b^k \in B(w^k)$, $k \in \mathcal{K}$, *then there exists a* $\mathcal{K}^1 \subset \mathcal{K}$ *such that*

$$b^k \longrightarrow b^\infty \quad k \in \mathcal{K}^1$$

where

$$b^\infty \neq 0$$

(d) *Both V and Y are compact.*

(e) *Either B is a continuous function and Y is compact or V is compact and C is a continuous function.*

The Import of Composition

To verify the closedness of A we shall instead prove that the maps forming A are closed; then when the above results are applied, A itself will be closed. This process will be illustrated amply in Chap. 5.

EXERCISES

4.1. Consider the QP problem

$$\max \quad q^t x + \tfrac{1}{2} x^t Q x$$
$$\text{subject to} \quad Ax = b \qquad (4.15)$$
$$x \geq 0$$

Assume the problem has a solution.

(a) Show that the K-T conditions for this problem are linear except for K-T condition (2) which is

$$\lambda_i x_i = 0$$
$$\lambda_i \geq 0 \qquad i = 1, \ldots, n$$

(b) Show how to solve the problem (4.15) by solving at most 2^n systems of linear equations.

Hint: Solve K-T conditions (1) and (3) with the additional provision that $\lambda_i = 0$, $i = 1, \ldots, n$. If the system has a solution with $x_i \geq 0$, $i = 1, \ldots n$, then K-T condition (2) also holds, and the resulting solution satisfies the K-T conditions. Secondly, solve K-T conditions (1) and (3) subject to $x_1 = 0$, $\lambda_i = 0$, $i = 2, \ldots, n$. If the solution yields $x_i \geq 0$, $i = 2, \ldots, n$, and $\lambda_1 \geq 0$ K-T condition (2) is also solved. Continue solving equations in this manner until all 2^n combinations of either $\lambda_i = 0$ or $x_i = 0$ have been tried.

Evaluate the objective function for all x's that solved the K-T conditions. The x that yields the largest objective function value is optimal. (Recall with linear constraints the constraint qualification holds.)

(c) Show that neglecting the round-off error the method for solving problem (4.15) described in (b) involves only a finite number of computer operations.

4.2. *An Equivalent Definition of Closed Map* (VEINOTT). Define the *graph* of a point-to-set map $A: U \longrightarrow V$ to be the set

$$\{(u, v) \,|\, u \in U, v \in A(u)\}$$

Prove that the following definition of closedness is equivalent to that in the text.

A point-to-set map $A: U \longrightarrow V$ is called closed if for each closed subset $W \subset U$, the graph $A: W \longrightarrow V$ is closed.

4.3. Prove that the point-to-set map in algorithm (4.4)

$$A(z) = \{y \,|\, \tfrac{1}{4}z \leq y \leq \tfrac{1}{2}z\}$$

is closed. Graph the map.

4.4. Let $A: U \longrightarrow V$ be a function. Suppose when A is interpreted as a point-to-set map (where the set is a single point) that A is closed at z^∞. Prove that if V is compact then the function A is continuous at z^∞. (See exercise 4.8 for why compactness is needed.)

4.5. Show that the below sets X are not compact
  (a) $X = \{x \,|\, 0 < x \leq 1, x \in E^1\}$
  (b) $X = E^n$

4.6. Which of the following are closed maps
   (a) $A(x) = \{y \mid y^t x \le b\}$ where x and y are n vectors and b is a scalar.
   (b) $A(x) = \{y \mid y^t x = b, \ y \ge 0\}$
   (c) Let $h_1(x), \ldots, h_l(x)$ be continuous functions from E^{n_1} to E^{n_2}. Consider $A: E^{n_1} \to E^{n_2}$ where

$$A(x) = \{y \in E^{n_2} \mid y^t h_i(x) \le b_i \qquad i = 1, \ldots, l\}$$

   (d)

$$A(x) = \begin{cases} \{y \mid y \le 2x\} & x \le 0 \\ \{y \mid y > 2x\} & x > 0 \end{cases}$$

   where $x, y \in E^1$.

4.7. Let $P(w)$ be an $m \times n$ matrix with continuous components $p_{ij}(w)$. Also let $b(w)$ be an $m \times 1$ column vector with continuous components. Define the map $B: W \to Y$

$$B(w) = \{y \mid P(w)y = b(w)\}$$

Thus $B(w)$ is the set of all solutions y to the equations $P(w)y = b(w)$. Show that $B(w)$ is a closed map.

4.8. *A Composition That Does Not Satisfy the Hypotheses of Lemma 4.2.* Let $B: E^1 \to E^1$ and $C: E^1 \to E^1$ be point-to-set maps (actually functions) defined as follows:

$$B(w) = \begin{cases} \left\{x \mid x = \dfrac{1}{w}\right\} & \text{if} \quad w \ne 0 \\ \{x \mid x = w\} & \text{if} \quad w = 0 \end{cases}$$

and

$$C(w) = \begin{cases} \left\{x \mid x = \dfrac{1}{w}\right\} & \text{if} \quad w \ne 0 \\ 1 & \text{if} \quad w = 0 \end{cases}$$

   (a) Prove B and C are closed maps.
   Note that the definition of closed map requires both $w^k \to w^\infty$ and $x^k \to x^\infty$, $k \in \mathscr{K}$. At the point $w^\infty = 0$ maps B and C are closed because $\lim_{w^k \to 0} (1/w^k)$ does not exist.
   (b) Let the map $A = BC$. Then

$$A(w) = \begin{cases} \{w\} & \text{if} \quad w \ne 0 \\ 1 & \text{if} \quad w = 0 \end{cases}$$

Show that A is not closed.

(c) Verify that the assumptions of Lemma 4.2 are not satisfied.

4.9. Prove that the composition of two continuous functions being itself continuous is a special case of Lemma 4.2.

4.10. Prove Corollaries 4.2.1 and 4.2.2.

4.11. Prove Lemmas 4.4, 4.5, and 4.6.

Hint: Consider Lemma 4.6 (b). Let $w^k \rightarrow w^\infty$ and $y^k \rightarrow y^\infty$ $k \in \mathcal{K}$ where $y^k \in A(w^k)$. Selecting the corresponding c^k, $c^k \rightarrow c^\infty$ $k \in \mathcal{K}^1$. Say $c^\infty = (c_i^\infty)$ and $c_1^\infty \neq 0$. Then

$$b^k \longrightarrow \frac{y_1^\infty}{c_1^\infty} \equiv b^\infty \qquad k \in \mathcal{K}^1$$

Moreover,

$$b^k c^k \longrightarrow b^\infty c^\infty = y^\infty$$

NOTES AND REFERENCES

The development of the algorithm and convergence is based upon ZANGWILL (1966g). Dr. J. Abadie has indicated to this author that work on the theory of convergence somewhat like Convergence Theorem A has been done by CHEVASSUS. The value to NLP of decomposing closed maps was indicated in ZANGWILL (1967d).

For further treatment of the mathematical properties of closed maps, see BERGE, which also reveals the intimate relationship between closed maps and upper semicontinuous maps. Discussions of compactness may be found in standard texts, such as APOSTLE; J. L. KELLEY; and SIMMONS. Figure 4.3 was suggested by Dr. A. C. Williams.

5 Unconstrained

Problems

Some of the most interesting applications of Convergence Theorem A and of the theorems on composition of closed maps are to algorithms for the unconstrained problem

$$\max \quad f(x) \qquad\qquad (5.1)$$

$$\text{over} \quad x \in E^n$$

This chapter presents several of these algorithms, each of which is a method of feasible directions. For these algorithms the form of the algorithmic map A is

$$A = M^1 D$$

Roughly, after the map D determines a direction, the map M^1 optimizes the objective function f in that direction. First we will study the map M^1 and prove it to be closed. Then once the appropriate map D is proved closed, A's closedness will follow easily by the theorems in Chap. 4. Indeed, the proofs in this chapter almost reduce to verifying that D is closed.

The algorithms to be studied in this chapter are the Cauchy steepest-ascent method, a modified Newton method, a cyclic-coordinate-ascent method, and a second-order method. In addition, the appendix considers the problem of one-dimensional search posed by the map M^1 and presents the golden-section and bisecting searches.

5.1. THE SOLUTION POINT DEPENDS UPON THE ALGORITHM

For all methods a certain difficulty arises due to their capability. Most, although not all, of the algorithms for problem (5.1) can only guarantee to determine a point x such that $f(x) = 0$. For such cases Ω, the solution set, is defined to be

$$\Omega = \{x \mid \nabla f(x) = 0\}$$

Unless we assume concavity or some related property, the condition $\nabla f(x) = 0$ may not imply that x is optimal. There is no value in defining Ω to be the set of optimal points if the algorithm cannot guarantee to achieve such a point, because we could not prove convergence.

The second-order method is more powerful than the other methods for it will determine a point x at which not only is $\nabla f(x) = 0$ but also certain second-order conditions hold. The set Ω will be defined accordingly. Nevertheless Ω will not be defined as the set of optimal points.

In general we define Ω to ensure that the algorithm can converge to a point in it. Thus the solution set Ω will depend upon the algorithm itself.

5.2. FEASIBLE DIRECTION METHODS AND
THE MAPS M^1 AND D

As mentioned above this chapter presents only feasible direction methods. Such algorithms iterate as follows. From a point x^k use a map D to determine a direction d^k. Then applying a map M^1 maximize the objective function on a segment either of the ray emanating from x^k in the direction d^k or of the line through x^k in the direction d^k. Call a point that achieves this one-dimensional maximization x^{k+1}. Mathematically

$$x^{k+1} = x^k + \tau^k d^k$$

where

$$f(x^{k+1}) = \max \quad \{f(x^k + \tau d^k) \,|\, \alpha \geq \tau \geq \beta\}$$

and α is either $+\infty$ or a positive scalar and, $\beta = 0$, $-\alpha$, or $-\infty$.

Let us now define the maps more precisely. The map $D: E^n \longrightarrow E^{2n}$ given x yields the point $(x, d) \in E^{2n}$, where d is a direction in E^n. The exact determination of the direction d will depend upon the algorithm.

The map $M^1: E^{2n} \longrightarrow E^n$ has the same form for all the algorithms in this chapter and is defined as follows:

$$M^1(x, d) = \{y \,|\, f(y) = \max_{\tau \in J} \; f(x + \tau d), y = x + \tau^0 d\}$$

where (x, d) is a point in E^{2n}, and J is an interval over which the scalar τ varies. Specific techniques for calculating $y \in M^1(x, d)$ will be given in the appendix to this chapter.

Using the definition of map composition given in Chap. 4 the algorithmic maps in this chapter will have the form

$$A = M^1 D$$

or will be compositions of this form.

5.3. CLOSEDNESS OF THE MAP M¹

Recall from Chap. 4 that in order to prove convergence we must establish the closedness of map A. However, if it is verified that both M^1 and D are closed, then after applying the theorems of Chap. 4, we know A will be closed. We now establish that the map M^1 is closed.

Lemma 5.1: *Let f be a continuous function. Then M^1 is closed if J is a closed and bounded interval.*

PROOF: Let
(a) $(x^k, d^k) \longrightarrow (x^\infty, d^\infty)$ $k \in \mathcal{K}$
(b) $y^k \in M^1(x^k, d^k)$ $k \in \mathcal{K}$
and
(c) $y^k \longrightarrow y^\infty$ $k \in \mathcal{K}$
Write

$$y^k = x^k + \tau^k d^k \tag{5.2}$$

where τ^k is an optimizing τ.

Because $\tau^k \in J, k \in \mathcal{K}$, and because J is closed and bounded, and hence, compact, there must be a convergent subsequence

$$\tau^k \longrightarrow \tau^\infty \quad k \in \mathcal{K}^1$$

where $\mathcal{K}^1 \subset \mathcal{K}$ and τ^∞ is in J.

Now for any fixed $\tau \in J$, by the definition of y^k

$$f(y^k) \geq f(x^k + \tau d^k)$$

and exploiting the continuity of f, we may take limits of the above equation obtaining

$$f(y^\infty) = \lim_{k \in \mathcal{K}^1} f(y^k) \geq \lim_{k \in \mathcal{K}^1} f(x^k + \tau d^k) = f(x^\infty + \tau d^\infty) \tag{5.3}$$

But as (5.3) holds for any $\tau \in J$, given $y^* \in M^1(x^\infty, d^\infty)$,

$$f(y^\infty) \geq f(y^*) \tag{5.4}$$

On the other hand, because $y^* \in M^1(x^\infty, d^\infty)$ maximizes f over all $\tau \in J$, and as $y^\infty = x^\infty + \tau^\infty d^\infty$, where $\tau^\infty \in J$,

$$f(y^*) \geq f(y^\infty) \tag{5.5}$$

Via (5.4) and (5.5)

$$y^{\infty} \in M^1(x^{\infty}, d^{\infty}) \quad \blacklozenge$$

The lemma greatly simplifies proving that the map $A = M^1 D$ is closed. Basically, we need only prove that D is closed and then apply the results of Chap. 4 to establish the closedness of A.

Example of M¹

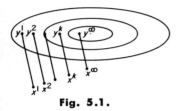

Fig. 5.1.

An example of Lemma 5.1 when $d^k = d$ for all k, is depicted in Fig. 5.1. Here $J = [0, \alpha]$ for α extremely large. The ovals indicate level lines of the function, and $y^k \in M^1(x^k, d)$. Certainly $y^{\infty} \in M(x^{\infty}, d)$.

Boundedness of J Required†

The assumption that the interval J was bounded played an important role in the lemma. Consider a second example similar to the above example, except that $J = [0, +\infty)$ so that J is not bounded. Let the direction be $d^1 = d$ and

$$d^k = \tfrac{1}{2}d^{k-1} = (\tfrac{1}{2})^{k-1}d$$

For this example the points $y^k \in M^1(x^k, d^k)$ are the same as in Fig. 5.1 because the two problems

$$f(x^k + \tau^0 d) = \max_{\tau \geq 0} \ f(x^k + \tau d)$$

and

$$f(x^k + \tau^k d^k) = \max_{\tau \geq 0} \ f(x^k + \tau d^k)$$

have the same answer

$$y^k = x^k + \tau^0 d = x^k + \tau^k d^k$$

---

†This part may be omitted on a first reading without loss of continuity. The reader may proceed immediately to Sec. 5.4.

where

$$\tau^0 = \tau^k (\tfrac{1}{2})^{k-1}$$

For both examples the points y^∞ are the same. However, at x^∞, for the second example

$$d^\infty = 0$$

so that

$$\max_{\tau \geq 0} \ f(x^\infty + \tau d^\infty) = \max_{\tau \geq 0} \ f(x^\infty)$$

and

$$y^\infty \notin M^1(x^\infty, d^\infty)$$

Consequently, if J is not bounded M^1 may not be closed.

For a third example, suppose that again $d^k = (\tfrac{1}{2})^{k-1} d$ but that J is taken to be bounded, say

$$J = [0, \alpha]$$

Then instead of Fig. 5.1 we have Fig. 5.2. In this case M^1 is closed at x^∞. The boundedness of J corrects for the $d^k \to 0$. Intuitively the length of the interval $[x^k + \alpha d^k, x^k]$ shrinks to zero because $d^k \to 0$. Thus even if $d^\infty = 0$, there is no difficulty (exercise 5.5).

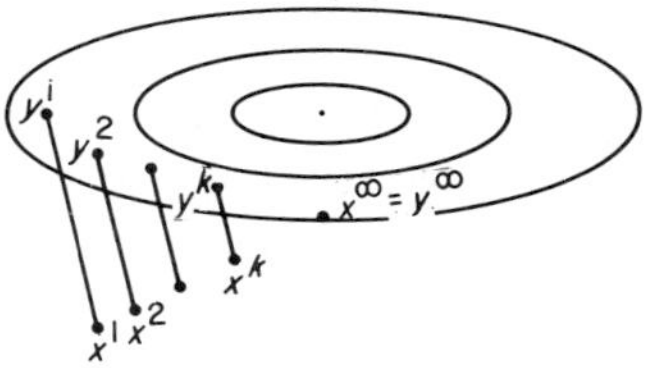

Fig. 5.2.

Thus we see the important role played by the boundedness of J. (This role is analogous to the role played by the assumption of Lemma 4.2 that ensured appropriate behavior of the sequences.) However a means for avoiding boundedness of J is suggested in exercises 5.5 and 5.6.

With Lemma 5.1 providing for the closedness of M^1, we now investigate various algorithms in order to specify the appropriate maps, prove that D is closed, and establish algorithmic convergence.

5.4. ALGORITHMS FOR THE UNCONSTRAINED PROBLEM

In the subsequent algorithms we assume all the points generated are on a compact set X. Almost invariably this is the case, and assuming it will

automatically establish condition 1 of Convergence Theorem A. Situations
for which this assumption does not hold will be considered later in the text.

Essentially the proofs will require verification of only two conditions.
First the direction d must produce an increase to f if we are not at a solution. Second, the map D must be closed. Then, because M^1 is closed, Convergence Theorem A will follow easily. In the next algorithm the verification
of these two conditions is almost immediate.

5.4.1. The Cauchy Steepest-Ascent Algorithm

In Theorem 2.1 it was observed that if $\nabla f(x) \neq 0$, by selecting $\nabla f(x)$
as the direction to move, one could obtain an increase in the objective function. The Cauchy procedure epitomizes this concept. Intuitively, at a point
x^k we calculate the gradient, and then find the point x^{k+1} by maximizing f
in the direction of the gradient from x^k.

The algorithmic map for the Cauchy procedure is

$$A = M^1 D^1$$

where $D^1 \colon E^n \rightarrow E^{2n}$ is the function

$$D^1(x) = (x, \nabla f(x))$$

and the interval $J = [0, \alpha]$ with α positive but finite. Often we select α sufficiently large to ensure that the maximum in M^1 occurs for some $\tau^0 < \alpha$,
although this is not necessary (see exercise 5.13).

Condition 2 will now be established. Define

$$z = x \qquad \text{and} \qquad Z(z) = f(x)$$

and call a point x a solution if

$$\nabla f(x) = 0$$

To verify 2(a) observe that if $\nabla f(x^k) \neq 0$ then by Theorem 2.1

$$\max_{\tau \in J} \ f(x^k + \tau \nabla f(x^k)) > f(x^k)$$

and thus

$$f(x^{k+1}) > f(x^k)$$

Condition 2(b) holds immediately.

We now attack condition 3. If f is assumed to have a continuous
derivative, $D^1(x)$ is a continuous function. Since the interval J is bounded,

Lemma 5.1 ensures that M^1 is a closed map. The results of Chap. 4 then verify that A is a closed map and condition 3 of Convergence Theorem A is proved.

With condition (1) assumed to hold, convergence has been established because the Cauchy procedure satisfies Convergence Theorem A.

5.4.2. The Modified Newton Procedure

The Cauchy procedure is founded upon linear, or first-order, Taylor approximations to f. Consider such an approximation f_L at x^1

$$f(x)_L = f(x^1) + \nabla f(x^1)^t(x - x^1)$$

The gradient is a good direction for maximizing the linear approximation, and because the linear approximation is quite accurate, especially in a neighborhood of x^1, the gradient direction is beneficial in obtaining an increase to the actual function f.

The Newton procedure extends the Cauchy notion one step further by exploiting second-order, or quadratic, approximations to f.

Behavior as Solution Point is Approached

Quadratic approximations not only are better approximations than linear approximations, but also become increasingly important as the solution point x^* is approached. A solution for the Newton procedure is defined as a point x^* at which $\nabla f(x^*) = 0$, just as in the Cauchy procedure. In terms of its Taylor expansion at a solution x^*, f becomes

$$f(x) = f(x^*) + (x - x^*)^t H(x^* + \theta(x - x^*))(x - x^*)$$

where $0 \leq \theta \leq 1$ and $H(x)$ is the $n \times n$ Hessian matrix of second partial derivatives evaluated at x. Observe that there is no linear term because at x^*, since x^* is a solution,

$$\nabla f(x^*) = 0$$

More generally, suppose we have a Taylor expansion at a point x' where x' is near x^*. When such an expansion is examined, since $\nabla f(x')$ is close to zero, the second-order terms often tend to dominate. Furthermore, in the neighborhood of x^* the individual components of the gradient, because they are near zero, may oscillate positive and negative. The geometric direction indicated by the gradient thus may change very rapidly. Basically, then, near the optimal point linear approximations behave poorly, and quadratic approximations are more stable.

Quadratic Approximations

Newton's procedure typifies techniques that exploit quadratic approximations. To motivate the method, let f_Q be a quadratic approximation to f at x^1.

$$f_Q(x) = f(x^1) + \nabla f(x^1)^t (x - x^1) + \tfrac{1}{2}(x - x^1)^t H(x^1)(x - x^1) \qquad (5.6)$$

If, f_Q were concave, then a necessary and sufficient condition for x^* to maximize f_Q over E^n is that

$$\nabla f_Q(x^*) = 0$$

or taking derivatives of (5.6)

$$\nabla f(x^1) + H(x^1)(x^* - x^1) = 0$$

Then assuming the matrix $H(x^1)$ is invertible

$$(x^* - x^1) = -H^{-1}(x^1)\nabla f(x^1) \qquad (5.7)$$

Observe that the direction vector $-H^{-1}(x^1)\nabla f(x^1)$ points from x^1 in the direction of that point x^* which optimizes the quadratic approximation to f.

In general f is not quadratic. However, because the right side of (5.6) is the quadratic approximation to f at x^1, the direction $-H^{-1}(x^1)\nabla f(x^1)$ is a good direction to seek a higher value of f. Newton's method utilizes this direction and the map M^1 to determine a better point.

The Newton procedure is thus identical to the Cauchy procedure except the direction $-H^{-1}(x)\nabla f(x)$ is used instead of $\nabla f(x)$. Specifically, the Newton method employs the direction map

$$D^2(x) = \left(x, -H(x)^{-1}\nabla f(x)\right)$$

where $H^{-1}(x)$ is the inverse of f's matrix of second partial derivatives evaluated at x. Then $A = M^1 D^2$. We further assume that $H^{-1}(x)$ is continuous and negative definite. By negative definite is meant

$$y^t H y < 0 \quad \text{for any} \quad y \neq 0$$

As the proof of convergence is analogous to the Cauchy proof, it is left as an exercise (exercise 5.1).

5.4.3. Cyclic Coordinate Ascent

The Cauchy procedure is useful if the first partial derivatives of f are available; while should both the first and second partial derivatives be easily obtainable, the Newton procedure is quite powerful. In many applications, however, although the function might be known to possess well-behaved partial derivatives, it might be quite difficult or time consuming to explicitly calculate them. For example, Newton's technique requires not only the determination of the Hessian but also the Hessian's inversion. The following procedure does not require calculation of any derivatives and is quite simple to implement. Whenever derivatives are difficult to calculate, it is a very valuable technique.

Let the n unit coordinate directions be specified

$$c_1, \ldots, c_n$$

For example in E^3

$$c_1 = \begin{pmatrix} 1 \\ 0 \\ 0 \end{pmatrix} \qquad c_2 = \begin{pmatrix} 0 \\ 1 \\ 0 \end{pmatrix} \quad \text{and} \quad c_3 = \begin{pmatrix} 0 \\ 0 \\ 1 \end{pmatrix}$$

The cyclic-coordinate-ascent method optimizes cyclically in each of the coordinate directions. Given a point x^k, set $x^{k,1} = x^k$. The point $x^{k,2}$ is obtained by maximizing in the positive and negative direction c_1 from $x^{k,1}$. Then we calculate $x^{k,3}$ by optimizing, using both the positive and negative direction c_2, from $x^{k,2}$. After determining $x^{k,n}$, we utilize the positive and negative direction c_n to obtain $x^{k+1,1} = x^{k+1}$. The procedure then repeats itself from $x^{k+1,1}$. By negative direction c_i is meant $-c_i$.

The Algorithmic Map

Defining $D^{3,i}(x) = (x, c_i)$, we discover the algorithmic map is the composition of the $2n$ maps,

$$A(x) = M^1 D^{3,n} M^1 D^{3,n-1} \ldots, M^1 D^{3,2} M^1 D^{3,1}$$

Here for M^1 the interval $J = [-a, +a]$ where a is a positive scalar. This choice of J provides for a search in both the positive and negative directions.

Convergence

Any map $M^1 D^{3,i}$ is closed by Corollary 4.2.2 because $D^{3,i}$ is a continuous function and M^1 is a closed map. By assumption all points are in a compact set X. Thus for each i

$$M^1 D^{3,i}: X \longrightarrow X$$

Corollary 4.2.1 then verifies that the map A is closed as it is the composition of closed maps on compact sets, and condition 3 of Convergence Theorem A is verified.

To prove condition 2 suppose f has continuous first partial derivatives. Furthermore, assume for any x and direction c_i that there is a unique τ' for which

$$f(x + \tau' c_i) = \max_{\tau \in J} \; f(x + \tau c_i)$$

This uniqueness assumption eliminates flat spots.

Let $z = x$ and $Z(z) = f(x)$, and term a point x a solution if

$$\max_{\tau \in J} \; f(x + \tau c_i) = f(x) \qquad i = 1, \ldots, n$$

At such a point, $\nabla f(x) = 0$ (exercise 5.2).

Condition 2(a) holds easily because if x^k is not a solution

$$f(x^{k+1}) > f(x^k)$$

The hypotheses of Convergence Theorem A are then verified because condition 1 is assumed.

5.4.4. A Second-Order Optimum Technique

The previous three methods for problem (5.1) could guarantee to converge to points x such that

$$\nabla f(x) = 0$$

Essentially, they calculated points at which the first-order necessary conditions for x to be optimal were satisfied. As discussed in Chap. 2 there exist many situations at which $\nabla f(x) = 0$ but x is not optimal. The second-order method is designed to avoid some of these situations.

But before developing the technique that will converge to a point at which both the first- and second-order necessary conditions for an optimum hold, we must derive the second-order conditions.

Second-Order Conditions for Optimality

Lemma 5.2: *Suppose f has continuous second partial derivatives and assume that at a point x the Hessian matrix $H(x)$ is not negative semidefinite. Then there exists a point y such that*

$$f(y) > f(x)$$

PROOF: Because $H(x)$ is not negative semidefinite there is a direction d^1 such that

$$(d^1)^t H(x) d^1 > 0 \tag{5.8}$$

If we define a direction d in terms of d^1 by

$$d = \delta d^1$$

where

$$\delta = \begin{cases} +1 & \text{if} \quad \nabla f(x)^t d^1 > 0 \\ -1 & \text{if} \quad \nabla f(x)^t d^1 < 0 \\ \pm 1 & \text{if} \quad \nabla f(x)^t d^1 = 0 \end{cases}$$

(If $\nabla f(x)^t d = 0$, choose $\delta = +1$ or -1 arbitrarily.) then

$$\nabla f(x)^t d \geq 0 \tag{5.9}$$

Using a Taylor expansion, we find

$$f(x + \tau d) = f(x) + \tau \nabla f(x)^t d + \tfrac{1}{2}\tau^2 d^t H(x + \theta \tau d) d \tag{5.10}$$

for some θ, $0 \leq \theta \leq 1$.

Due to Equation (5.8) and the fact that H is continuous, there must be a $\tau = \tau^0 > 0$ sufficiently small that

$$d^t H(x + \theta \tau^0 d) d > 0 \tag{5.11}$$

Letting $y = x + \tau^0 d$, and incorporating (5.9) and (5.11) into Equation (5.10)

$$f(y) = f(x) + \tau^0 \nabla f(x)^t d + \tfrac{1}{2}(\tau^0)^2 d^t H(x + \theta \tau^0 d) d > f(x) \quad \blacklozenge$$

Lemma 5.2 actually describes how to obtain an increased value of f should $H(x)$ not be negative semidefinite. The second-order necessary optimum conditions, Theorem 5.3, follow immediately.

Theorem 5.3: *Let f possess continuous second partial derivatives. Suppose x^* maximizes f over E^n. Then*

$$\nabla f(x^*) = 0$$

and the Hessian matrix

$$H(x^*)$$

is negative semidefinite.

PROOF: That $\nabla f(x^*) = 0$ follows from Corollary 2.1.1. Lemma 5.2 provides the second-order condition that $H(x^*)$ be negative semidefinite. ◆

Eigenvector Properties

As might be expected from the hypotheses of Theorem 5.3, we will assume in this section that f has continuous second partial derivatives. But before developing a technique that will seek points satisfying both the first- and second-order conditions, we must recall some facts proved in calculus and linear algebra.

Since f has continuous second partial derivatives,

$$\text{the Hessian matrix } H(x) \text{ is a symmetric } n \times n \text{ matrix} \qquad (5.12)$$

Any symmetric $n \times n$ matrix H has n eigenvalues λ_i, $i = 1, \ldots, n$, (some of which may be equal) and n unit eigenvectors e_i, $i = 1, \ldots, n$.† These eigenvectors and eigenvalues have many properties, some of which we now summarize.

Corresponding to a given eigenvalue, say λ_i, there is a unit eigenvector e_i that satisfies the equation

$$He_i = \lambda_i e_i \qquad (5.13)$$

$$\|e_i\| = 1$$

The unit eigenvectors thus found are orthonormal, that is

$$e_i^t e_j = 0 \qquad i \neq j$$

and

$$\|e_i\| = 1 \qquad (5.14)$$

where $\|\cdot\|$ denotes Euclidean norm and $\|e_i\|^2 = e_i^t e_i$.
In terms of its eigenvectors and eigenvalues, the matrix H has a diagonal representation

$$H = E^t \Lambda E \qquad (5.15)$$

---

†Eigenvectors and eigenvalues are often referred to as characteristic vectors and characteristic values, respectively.

where E is a matrix whose rows are the n eigenvectors

$$E = \begin{bmatrix} e_1^t \\ e_2^t \\ \cdot \\ \cdot \\ \cdot \\ e_n^t \end{bmatrix} \tag{5.16}$$

and Λ is a diagonal matrix with the n eigenvalues on the diagonal

$$\Lambda = \text{diag} \ [\lambda_1, \lambda_2, \ldots, \lambda_n] \tag{5.17}$$

It follows from the representation of H in (5.15) and the properties of the eigenvectors in (5.14) that

$$(e_i)^t H e_i = e_i^t E^t \Lambda E e_i = \lambda_i \tag{5.18}$$

Extending this reasoning, we can show that the matrix H is negative semi-definite if and only if

$$0 \geq \lambda_i \qquad i = 1, \ldots, n \tag{5.19}$$

(exercise 5.4).

Since the Hessian matrix $H(x)$ depends upon x, we illustrate this dependence explicitly for $H(x)$'s eigenvectors and eigenvalues

$$\lambda_i(x) \quad \text{and} \quad e_i(x)$$

Furthermore, we assume that the eigenvalues are ordered

$$\lambda_1(x) \geq \lambda_2(x) \geq \ldots \geq \lambda_n(x) \tag{5.20}$$

Hence, $\lambda_1(x)$ represents the largest eigenvalue and $e_1(x)$ an eigenvector corresponding to $\lambda_1(x)$.

Because f has continuous second partial derivatives, each element of $H(x)$ is a continuous function of x. The calculus then provides that

$$\lambda_1(x) \quad \text{is itself a continuous function of} \quad x \tag{5.21}$$

Thus at each x if we calculate the largest eigenvalue $\lambda_1(x)$ we determine a continuous function.

The Second-Order Algorithm

Having recalled these facts we develop the second-order procedure. Let $z = x$ and $Z(z) = f(x)$. The algorithmic map has the form

$$A = M^1 D^4$$

where in M^1 the interval $J = [0, a]$ for a positive. The map $D^4(x)$ takes x into (x, d) where

$$d = \nabla f(x) + \delta(x) e^*(x)$$

$$\delta(x) = \begin{cases} +1 & \text{if} \quad \nabla f(x)^t e^*(x) > 0 \\ -1 & \text{if} \quad \nabla f(x)^t e^*(x) < 0 \\ \pm 1 & \text{if} \quad \nabla f(x)^t e^*(x) = 0 \end{cases}$$

and $e^*(x)$ is the vector

$$e^*(x) = \max \quad [\lambda_1(x), 0] \, e_1(x)$$

Should $\nabla f(x)^t e^*(x) = 0$, $\delta(x)$ may be $+1$ or -1 arbitrarily. Here, of course, $\lambda_1(x)$ is the largest eigenvalue, and $e_1(x)$ is a corresponding eigenvector.

The direction d has two parts. The first part is the gradient and its role is the same as in the Cauchy procedure. In the second part, $e^*(x)$ is designed to take advantage of any lack in negative semidefiniteness of the matrix $H(x)$. Observe that $e^*(x) \neq 0$ only when $H(x)$ is not negative semidefinite. This follows by (5.19) because if $H(x)$ is negative semidefinite then $0 \geq \lambda_1(x)$. The $\delta(x)$ factor is analogous to the δ of Lemma 5.2 and ensures

$$\nabla f(x)^t \delta(x) e^*(x) \geq 0 \tag{5.22}$$

Proof of Convergence

We call a point x a solution if both

$$\nabla f(x) = 0$$

and

$$H(x) \text{ is negative semidefinite.}$$

Thus the solution points will satisfy both the first- and second-order necessary conditions.

Convergence will now be proved via Convergence Theorem A. Recall condition 1 is assumed to hold.

Condition 2 of Convergence Theorem A will now be established. Let x not be a solution. Suppose in the first case that

$$\nabla f(x) \neq 0$$

Then, using Equation (5.22)

$$\nabla f(x)^t d = \nabla f(x)^t \nabla f(x) + \nabla f(x)^t \delta(x) e^*(x) > 0$$

and Theorem 2.1 provides the result.

In the second case where x is not a solution $\nabla f(x) = 0$, but $H(x)$ is not negative semidefinite. Here

$$d = \delta(x) \lambda_1(x) e_1(x)$$

where $\delta(x) = +1$ or -1 and $\lambda_1(x) > 0$ because $H(x)$ is not negative semidefinite.

Then

$$d^t H(x) d = \delta(x)^2 \lambda_1(x)^2 e_1^t H(x) e_1$$

and from Equation (5.18)

$$= \lambda_1(x)^2 \lambda_1(x)$$
$$> 0 \quad \text{as} \quad \lambda_1(x) > 0$$

In addition by (5.22),

$$\nabla f(x)^t d \geq 0$$

An improvement can then be obtained using the process shown in the proof of Lemma 5.2 and condition 2(a) is established. Condition 2(b) holds immediately.

Proof of Condition 3

To demonstrate that the procedure converges only condition 3 remains. The map D^4 will be dissected into its components and each one proved to be closed. Then the results of Chap. 4 will establish A's closedness.

Clearly D^4 can be decomposed as follows. First the map $\mathscr{E}^1(x) = \{e_1(x)\}$ obtains the eigenvector $e_1(x)$ from x. Then the map $\mathscr{E}^2(x) = \{e^*(x)\}$ is simply the scalar-vector product of the function max $[\lambda_1(x), 0]$ and the map $\mathscr{E}^1$. By defining the map $\Delta(x) = \{\delta(x)\}$, the map $\mathscr{E}^3(x) = \{\delta(x)e^*(x)\}$ is as the scalar-vector product of Δ and $\mathscr{E}^2$. Finally by taking the sum of $\nabla f(x)$ and $\mathscr{E}^3(x)$, we obtain D^4. Let us now prove that these maps are closed.

To begin consider the map $\mathscr{E}^1$ that given x yields $e_1(x)$

$$\mathscr{E}^1(x) = \{e_1(x)\}$$

Let

$$x^k \longrightarrow x^\infty \qquad k \in \mathscr{K}$$

and

$$e_1(x^k) \longrightarrow e_1^\infty \qquad k \in \mathscr{K}$$

It must be shown that

$$e_1^\infty \in \mathscr{E}^1(x^\infty)$$

In other words, it must be established that e_1^∞ is a unit eigenvector for $\lambda_1(x^\infty)$. The eigenvector $e_1(x^k)$ solves Equation (5.13), which may be rewritten as

$$(H(x^k) - \lambda_1(x^k)I)e_1(x^k) = 0 \tag{5.23}$$

where I is the identity matrix and

$$\lambda_1(x^k)I = \text{diag} \quad (\lambda_1(x^k), \lambda_1(x^k), \ldots, \lambda_1(x^k))$$

Recalling that both $H(x^k)$ and $\lambda_1(x^k)$ are continuous and taking limits of Equation (5.23), we arrive at

$$(H(x^\infty) - \lambda_1(x^\infty)I)e_1^\infty = 0$$

Hence, e_1^∞ is itself an eigenvector for $\lambda_1(x^\infty)$. In addition all $e_1(x^k)$ are normalized, so that

$$\|e_1(x^k)\| = 1$$

and when the limit is taken

$$\|e_1^\infty\| = 1$$

Consequently, e_1^∞ is a unit eigenvector for $\lambda_1(x^\infty)$, and

$$e_1^\infty \in \mathscr{E}^1(x^\infty)$$

Thus, the map $\mathscr{E}^1$ is closed.

Now let us analyze the map $\mathscr{E}^2$ where

$$\mathscr{E}^2(x) = \{\max [\lambda_1(x), 0] \, e_1 \, (x) = e^*(x)\}$$

This map can be expressed as the scalar-vector product of two closed maps, one map being $\mathscr{E}^1$ and the other the continuous function $\max [\lambda_1(x), 0]$.

Because from (5.14) $\|e_1(x)\| = 1$, the $e_1(x)$ are contained in a closed and bounded, and hence compact, set. Lemma 4.6e applies, and $\mathscr{E}^2$ is closed.

In our construction of the map D^4, we now must prove closedness of the map

$$\Delta(x) = \{\delta(x)\}$$

This map, given x, determines $\delta(x)$.

Let

$$x^k \longrightarrow x^\infty \qquad x \in \mathscr{H}$$

and

$$\delta(x^k) \longrightarrow \delta^\infty \qquad x \in \mathscr{H} \tag{5.24}$$

Closedness will be proved by demonstrating that $\delta^\infty \in \Delta(x^\infty)$, or in other words, $\delta^\infty = \delta(x^\infty)$.

Because the $e_1(x)$ are on a compact set and max $[\lambda_1(x), 0]$ is a continuous function, there must be a $\mathscr{H}^1 \subset \mathscr{H}$ such that

$$e*(x^k) \longrightarrow e*(x^\infty) \qquad k \in \mathscr{H}^1 \tag{5.25}$$

If, in the first case,

$$\nabla f(x^\infty)'e*(x^\infty) > 0 \tag{5.26}$$

then employing (5.25) and the continuity of the gradient, for k sufficiently large and $k \in \mathscr{H}^1$

$$\nabla f(x^k)'e*(x^k) > 0$$

But by definition of $\delta(x^k)$, for these k

$$\delta(x^k) = +1$$

then from (5.24)

$$\delta^\infty = 1$$

Yet (5.26) provides

$$\delta(x^\infty) = 1$$

Therefore $\delta^\infty = \delta(x^\infty)$, and in this case Δ is closed.

Should $\nabla f(x^\infty)'e*(x^\infty) < 0$, the reasoning is similar.

If $\nabla f(x^\infty)'e^*(x^\infty) = 0$, then $\delta(x^\infty) = \pm 1$ arbitrarily, so that $\delta^\infty = \delta(x^\infty)$ must hold.

Thus all three cases are verified, and Δ is closed.

Our construction is up to the map

$$\mathscr{E}^3(x) = \{\delta(x)e^*(x)\} \tag{5.27}$$

But this is the scalar-vector product of the maps Δ and $\mathscr{E}^2$. Closedness of $\mathscr{E}^3$ follows by Lemma 4.6c as δ is in a compact set and $\delta \neq 0$.

Finally the map D^4 is the sum of the map $\nabla f(x)$ and $\mathscr{E}^3(x)$. Since ∇f is continuous and $\mathscr{E}^3$ is closed, Lemma 4.4 proves that D^4 must be closed.

To prove the closedness of $A = M^1 D^4$, we apply Lemma 4.2. It is left as exercise 5.7 to show that the hypotheses of Lemma 4.2 are satisfied so that the map A is closed.

Consequently, the second-order method converges.

APPENDIX: ONE-DIMENSIONAL-SEARCH PROCEDURES

All of the algorithms for problem (5.1) discussed in this and the next chapter embody the problem of determining

$$y \in M^1(x, d)$$

This latter problem of one-dimensional search will now be discussed.

5.A.1. BACKGROUND

One-dimensional searches attempt to determine a point w^*, called optimal, that maximizes f over an interval L. For general continuous functions this problem is quite difficult, and *ad hoc* procedures are necessary. Usually the problem requires an exhaustive search of the interval, and even then there could be difficulties.

Under the additional assumption of concavity, however, there are powerful and efficient procedures, and we will now discuss two of them.

Our discussion assumes that all points are in E^1. Even though the points for problem (5.1) are actually in E^n, there is no loss in generality by this assumption because the search procedure operates as if it were in one dimension. The procedures start with an interval $L^1 = L = [u^1, v^1]$ containing the optimal point w^*. Then, given an interval

$$L^k = [u^k, v^k]$$

containing w^*, they determine

$$L^{k+1} = [u^{k+1}, v^{k+1}]$$

also containing w^* such that, in addition, L^{k+1} is strictly contained in L^k. Always

$$u^k \leq u^{k+1} \leq w^* \leq v^{k+1} \leq v^k$$

and either $u^k < u^{k+1}$ or $v^{k+1} < v^k$. The intervals thereby get smaller and smaller until w^* is reached in the limit.

5.A.2. THE GOLDEN-SECTION OR FIBONACCI SEARCH

The golden-section search is for f concave. It requires use of the Fibonacci fractions

$$F_1 = \frac{3 - \sqrt{5}}{2} \approx 0.38$$

and

$$F_2 = \frac{\sqrt{5} - 1}{2} = 1 - F_1 \approx 0.62$$

Observe $F_1 = (F_2)^2$.

Given an interval $L = [u, v]$ of length $l = v - u$, the next interval is selected as follows. Let

$$w_1 = u + F_1 l$$

and

$$w_2 = u + F_2 l = v - F_1 l$$

be points on the interval.

If $f(w_1) > f(w_2)$,
    the new interval is $[u, w_2]$.
If $f(w_1) < f(w_2)$,
    the new interval is $[w_1, v]$.
If $f(w_1) = f(w_2)$,
    the new interval is either
    $[u, w_2]$ or $[w_1, v]$.

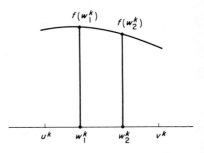

Fig. 5.3.

The golden-section procedure. Here $f(w_1^k) > f(w_2^k)$. Hence $u^{k+1} = u^k$, $v^{k+1} = w_2^k$.

We are assured by concavity that the optimal point w^* is on the resulting interval (see Fig. 5.3).

It should be clear that this process generates a sequence of intervals $\{L^k\}_1^\infty$ that in the limit closes in on the optimal point w^∞.

Efficiency of the Search

In the golden-section-search procedure $f(w_1)$ and $f(w_2)$ have to be compared. It therefore appears that we must make two evaluations, both $f(w_1)$ and $f(w_2)$, at each iteration. However, a key property of this procedure is that only one new point need be evaluated at each iteration due to the fact that one of the points has already been determined.

For example, suppose on iteration k that we have calculated w_1^k and w_2^k and that

$$f(w_1^k) > f(w_2^k)$$

Then $(u^{k+1}, v^{k+1}) = (u^k, w_2^k)$. Now w_1^{k+1} and w_2^{k+1} are to be evaluated. But (exercise 5.8) $w_2^{k+1} = w_1^k$ and $f(w_1^k)$ has already been determined. The Fibonacci fractions thus guarantee that f will be evaluated at only one new point.

5.A.3. THE BISECTING OR BOLZANO SEARCH

For this search f is assumed to be concave and to have a continuous derivative. The algorithmic map at each iteration divides the present interval in half. Depending upon the derivative at the midpoint, either the left half or the right half becomes the new interval. If the derivative at the midpoint indicates the optimal point is to the right of the midpoint, then the new interval is the right half, and analogously for the left. If the derivative is zero, then by concavity the midpoint is optimal.

In this manner the search generates intervals each one-half the length of its immediate predecessor. As each remaining interval must contain the optimal point, the bisecting search clearly converges.

EXERCISES

5.1. Prove convergence of the modified Newton procedure.

5.2. Prove in the cyclic-coordinate-ascent method that

$$\max_{\tau \in J} \ f(x + \tau c_i) = f(x) \qquad i = 1, \ldots, n$$

implies $\nabla f(x) = 0$ if there is a unique τ' such that

$$f(x + \tau' c_i) = \max_{\tau \in J} \ f(x + \tau c_i) \qquad i = 1, \ldots, n$$

Hint: Use the uniqueness to verify that

$$f(x + \tau c_i) < f(x) \quad \text{for} \quad -\alpha < \tau < 0 \quad \text{and} \quad 0 < \tau < \alpha$$

Hence

$$\frac{\partial f(x)}{\partial x_i} = 0$$

5.3. Prove convergence of the procedure for problem (5.1)
$$A = M^1 D^5$$
where $D^5(x) = (x, d)$ and

$$d = \frac{\partial f(x)}{\partial x_i}, \text{ where } \left| \frac{\partial f(x)}{\partial x_i} \right| = \max \ \left| \frac{\partial f(x)}{\partial x_j} \right|$$

What assumptions are needed?

5.4. Let H be a symmetric $n \times n$ matrix with eigenvalues λ_i, $i = 1, \ldots, n$, and eigenvectors e_i, $i = 1, \ldots, n$.
  (a) Show that any n vector x can be written as a linear combination of the eigenvectors e_i.
  Hint: use orthonormality.
  (b) Prove that H is negative semidefinite if and only if

$$0 \geq \lambda_i$$

5.5. Let f be a continuous function. Consider the map $M^1(x, d)$ where $J = [0, +\infty)$. Prove that if $d^\infty \neq 0$, then M^1 is closed at (x^∞, d^∞). Assume all maxima exist.
  Hint: Let $x^k \longrightarrow x^\infty$, $d^k \longrightarrow d^\infty$ $k \in \mathcal{K}$ where

$$x^{k+1} = x^k + \tau^k d^k$$

Use the fact that $d^\infty \neq 0$ to show that there is a $\mathcal{K}^1 \subset \mathcal{K}$ such that

$$\tau^k \longrightarrow \tau^\infty \qquad k \in \mathcal{K}^1$$

5.6. Consider those algorithms in the chapter that used the map M^1 with $J = [0, \alpha]$ for α very large. Prove that these algorithms still converge with $J = [0, +\infty)$.

Hint: The map A need only be closed at a point x not a solution. At such a point $d \neq 0$. Then use exercise 5.5 to establish that A is closed.

5.7. Prove in the second-order method that $A = M^1 D^4$ is a closed map using Lemma 4.2.

Hint: M^1 and D^4 are closed. Let $x^k \longrightarrow x^\infty$ $k \in \mathcal{K}$. Prove there is a $\mathcal{K}^1 \subset \mathcal{K}$ such that $d^k \longrightarrow d^\infty$. Use the facts that $\nabla f(x)$ and max $[\lambda_1(x), 0]$ are continuous functions and that $\delta(x)$ and $e_1(x)$ are on compact sets.

5.8. In the golden-section search show that if

   (a) $(u^{k+1}, v^{k+1}) = (u^k, w_2^k)$ then $w_1^k = w_2^{k+1}$

while if

   (b) $(u^{k+1}, v^{k+1}) = (w_1^k, v^k)$ then $w_2^k = w_1^{k+1}$

5.9. Suppose an initial interval L is length l. After k iterations what is the length of interval L^k in the (a) golden-section search, and (b) bisecting search? Which has reduced the interval more by the kth iteration? Which would you prefer to use on a computer recalling the computer effort involved in the evaluation of a directional derivative.

5.10. A function $f(x)$ for $x \in E^1$ is called strictly unimodal if it is continuously differentiable and

$$\nabla f(x) = 0 \quad \text{implies} \quad x \quad \text{maximizes} \quad f \quad \text{over} \quad E^1$$

A point x^* such that $\nabla f(x^*) = 0$ is termed optimal. Does the golden-section or bisecting search also converge for strictly unimodal functions?

5.11. Prove convergence of the golden-section search via Convergence Theorem A.

5.12. Prove that the bisecting search converges by applying Convergence Theorem A.

5.13. (GINSBURG, S.) Often in computer implementation calculation of the map M^1 is time consuming. Usually, the bisecting- and golden-section-search procedures produce rapid increases in the objective function during their first few iterations. However, after these initial iterations the final determination of the optimal point is often time consuming and yields little further increase to f. A method that is often more efficient employs a map that either optimizes f on the interval or increases f an amount at least $\epsilon > 0$. Thus we may stop the search whenever either f increases at least ϵ or if such an increase is impossible f is maximized on the interval. We define such a map as M^*

$$M^*(x, d) = M^1(x, d) \cup \{y = x + \tau d \,|\, f(y) \geq f(x) + \epsilon\}$$

where $\tau \in J$.

Prove that the map M^* is closed if the interval J is closed and bounded.

NOTES AND REFERENCES

§5.2. The terms feasible direction and methods of feasible direction were suggested by ZOUTENDIJK (1960).

For other forms of maps for feasible direction algorithms see exercise 5.13 as well as GOLDSTEIN (1965), (1966); and TOPKIS and VEINOTT.

§5.3. This section is based upon a paper by ZANGWILL (1967d).

§5.4. Sections 5.4.1 and 5.4.2 can be found in ZANGWILL (1967d).

Once it is understood that the gradient produces an increase and once the closedness theorems have been assimilated, then the proof of the Cauchy procedure reduces to one line. Specifically, the proof is simply that the gradient is a continuous function. Similar statements hold for the modified Newton and cyclic method.

Various references to the Cauchy method include CAUCHY; CURRY; FORSYTHE; GOLDSTEIN (1962); IVANOV (1962a); and KANTOROVICH and AKILOV. The modified Newton is based upon FIACCO and MCCORMICK (1963). For the cyclic method see D'ESOPO; SCHECTER; and IVANOV (1962b). The second-order method is a simplification of more general approaches found in MCCORMICK and ZANGWILL. However, it should be remarked that for all four algorithms the previous authors did not use the convergence conditions of Convergence Theorem A and generally required considerably more complicated proofs.

For an interesting excursion into the difficulties that can ensue when the objective function is not continuously differentiable, see ZUHOVICHII, POLJAK, and PRIMAK; and ZANGWILL (1967f).

The relation between first- and second-order methods is examined in CROCKETT and CHERNOFF.

For further discussion of eigenvectors (characteristic vectors), see standard algebra texts such as HADLEY (1961); HALMOS; BIRKHOFF and MAC-LANE; and FINKBEINER.

§5.A.2. For the original paper refer to PISANO as cited by COXETER. The Fibonacci search has certain "optimal" properties as a procedure and relates to the Fibonacci numbers. For a discussion of these aspects, see BELLMAN (1957) p. 34; and KEIFER. A related procedure is presented in OLIVER and WILDE.

§5.A.3. A thorough discussion of this search can be found in WILDE; and WILDE and BEIGHTLER. Indeed, many types of one-dimensional and other searches are discussed in these two references.

6 Mixed Algorithms and Quadratic Acceleration Techniques Via Conjugate Directions

Each of the four methods for the unconstrained problem, problem (5.1), discussed in the previous chapter possesses its own advantages and disadvantages. In this chapter we suggest that the algorithms be mixed. For example, when maximizing a given function, we might use one algorithm during that part of the optimization process for which it is most efficient, while another algorithm might be employed during that portion of the process for which it is most efficient. Speed of convergence should thus be enhanced. After defining a mixed algorithm we will validate this concept by proving Convergence Theorem B. Then we will present two mixed algorithms for accelerating convergence that exploit conjugate directions and quadratic approximations. Both algorithms will maximize a quadratic, $q^t x + \frac{1}{2} x^t Q x$, in a finite number of steps, although the second algorithm requires a negative definite Q.

6.1. MIXED ALGORITHMS FOR ENHANCING SPEED OF CONVERGENCE

Of the four previously presented methods for maximizing a function, the cyclic-coordinate-ascent method is undoubtedly the simplest to implement, as it requires no derivatives. Indeed, during certain stages of the maximization process derivatives might be extremely difficult to calculate, and only the cyclic method can be utilized. During other stages of the process, however, calculation of derivatives may be easier, and one of the other methods may be more efficient. As another example, suppose at the kth iteration that the Hessian $H(x^k)$ is negative definite; the Newton method could then be employed. However, if, at point x^{k+1}, $H(x^{k+1})$ has a positive eigenvalue, then the Newton method is useless, but the second-order method, because it exploits positive eigenvalues, may be helpful.

Convergence of a Mixed Algorithm

It is intuitively reasonable that as long as one of the methods in Chap. 5 is used at each iteration, no matter which method, then nothing should go wrong. In fact, suppose we use a given convergent procedure infinitely often, say during those iterations k for $k \in \mathcal{K}$. If during the other iterations the objective function is not allowed to decrease, then it seems reasonable that this algorithm will still determine an optimal point. Presumably during those other iterations we are employing some special technique in an effort to hasten convergence.

In order to state the theorem which validates convergence of these algorithms, observe that the algorithm might depend upon the points $z^1, \ldots, z^{k-1}$ as well as z^k. Hence, to indicate this dependence the algorithmic map will be denoted by A_k. Specifically, we define a **mixed algorithm** to be an algorithm that has a given basic algorithmic map B, which depends only upon z, such that

$$B = A_k \qquad k \in \mathcal{K}$$

In other words, the basic map B is used infinitely often. For the remaining k, other maps are employed.

In the theorem the basic map B will possess the properties of conditions 1, 2, and 3 of Convergence Theorem A. Hence, if $A_k = B$ for all k, the algorithm will converge by Convergence Theorem A. In general for the mixed algorithm, however, $A_k = B$ infinitely often but not for all k.

Convergence Theorem B: *Suppose there is an algorithmic map $B: V \longrightarrow V$ for the NLP problem (with associated Z function and solution set Ω) that satisfies conditions 1, 2, and 3 of Convergence Theorem A. Let a mixed algorithm for the problem be defined by the maps $A_k: V \longrightarrow V$ such that for some $\mathcal{K}$*

$$A_k = B \qquad k \in \mathcal{K}$$

while for $k \notin \mathcal{K}$

$$Z(z^{k+1}) \geq Z(z^k) \tag{6.1}$$

Further assume that
   (1) All $z^k \in X$ where X is compact, and
   (2) If $z^* \in \Omega$, and

$$Z(y) \geq Z(z^*)$$

then

$$y \in \Omega$$

   Then under these hypotheses the mixed algorithm either stops at a solution, or generates a sequence $\{z^k\}_1^\infty$ such that the limit of any convergent subsequence is a solution point.

   PROOF: Consider an arbitrary convergent subsequence $z^k \longrightarrow z', k \in \mathcal{K}'$. We must prove that z' is a solution. (Observe that $\mathcal{K}'$ need not be related to $\mathcal{K}$.)

By Equation (6.1) and the properties of map B

$$Z(z^{k+1}) \geq Z(z^k) \quad \text{for all} \quad k$$

and from Lemma 4.1

$$\lim_{k \to \infty} Z(z^k) = Z(z') \tag{6.2}$$

Condition 1 of this theorem provides that there is a $\mathcal{K}^1 \subset \mathcal{K}$ such that $z^k \longrightarrow z^\infty$, $k \in \mathcal{K}^1$. Again employing Lemma 4.1 we find that

$$\lim_{k \to \infty} Z(z^k) = Z(z^\infty) \tag{6.3}$$

Therefore, from (6.2) and (6.3)

$$Z(z') = Z(z^\infty) \tag{6.4}$$

By using the same reasoning as in the proof of Convergence Theorem A, since $A_k = B$ for $k \in \mathcal{K}^1$, we find that z^∞ must be a solution (exercise 6.1). Condition 2 of this theorem and (6.4) then provide the conclusion that

$$z' \in \Omega \quad \blacklozenge$$

Implications of the Theorem

In this chapter when we apply Convergence Theorem B to the unconstrained problem, $x = z, f = Z$, and all points generated are assumed to be on a compact set X. Convergence Theorem B merely requires that for some infinite subset $\mathcal{K}$, a convergent algorithm, say one of the four previously presented, be used on those iterations $k \in \mathcal{K}$. For the other iterations use any technique as long as

$$f(z^{k+1}) \geq f(z^k)$$

Then at the limit of any convergent subsequence $z^k \longrightarrow z^\infty$ $k \in \mathcal{K}'$, the objective function value $f(z^\infty)$ will be at least as large as the objective function value at a solution point for basic algorithmic map B.

6.2. CONJUGATE DIRECTIONS

Algorithms for the unconstrained problem that employ conjugate directions when the basic map B is not being used have been found to be quite efficient. Essentially, conjugate directions exploit quadratic approximations to f without explicitly calculating the Hessian matrix; rather, the requisite information about the Hessian is developed over a period of several iterations. Contrast this with the Newton-method approach of calculating the Hessian explicitly at each iteration. Intuitively, the Newton method, by recalculating the Hessian each time, obtains a better approximation to f than the conjugate-direction procedure. Yet it obtains this accuracy at the price of considerably more effort per iteration. Moreover, the first conjugate method does not require a negative definite Hessian.

Often we use conjugate directions in conjunction with either the Cauchy or cyclic-coordinate-ascent methods. Such combinations can be of great advantage because both the Cauchy and the cyclic method might converge only in a limit sense with a quadratic function. However, conjugate direction techniques will optimize a quadratic in a finite number of iterations. For general functions coupling either the Cauchy or Cyclic method with conjugate directions yields a powerful approach, because the resulting mixed algorithm will exploit quadratic approximations yet avoid extensive calculations of second partial derivatives.

Conjugate Directions

Given a $n \times n$ symmetric matrix Q, the directions $d_1, d_2, \ldots, d_r$, $r \leq n$, are said to be conjugate (or more precisely Q conjugate) if

$$\text{the} \quad d_i \quad \text{are linearly independent} \tag{6.5}$$

and

$$d_i^t Q d_j = 0 \qquad i \neq j \tag{6.6}$$

The next theorem suggests how conjugate directions can be used to maximize a quadratic. However, we must first discuss subspaces and manifolds.

Subspaces and Manifolds

Since r conjugate directions must be linearly independent, r conjugate directions span an r-dimensional subspace. In other words, the point

$$y = \sum_{i=1}^{r} \gamma_i d_i$$

generates an r-dimensional subspace as the γ_i vary, $-\infty < \gamma_i < +\infty$. Consider now the point y^1 where, given x^1 and $r \leq n$,

$$y^1 = x^1 + \sum_{i=1}^{r} \gamma_i d_i$$

In this case as the γ_i vary they generate an r-dimensional (linear) manifold.

An r-dimensional manifold is a translation of an r-dimensional subspace. Roughly, the x^1 in the above equation indicates the amount of translation from the origin, point 0. Note, the manifold may be a subspace if $x^1 = 0$ or if x^1 is in the subspace spanned by the d_i. However, if x^1 is not in the subspace spanned by the d_i, the manifold is no longer a subspace because a subspace must by definition contain the origin. Observe, however, that any n-dimensional manifold spans the entire space E^n. The n conjugate directions are linearly independent and consequently must span E^n.

In summary, a point x^1 and r conjugate directions generate an r-dimensional manifold, and any n-dimensional manifold is E^n.

In the next theorem we establish a key property of conjugate directions, viz., they maximize a quadratic function in a manifold.

Theorem 6.1: *Let*

$$p(x) = q^t x + \tfrac{1}{2} x^t Q x$$

be a quadratic function, and let $d_1, \ldots, d_r$ be conjugate. Suppose a point x' and directions $d_1, \ldots, d_r$ generate an r-dimensional manifold. Then, by searching the positive and negative directions $d_1, \ldots, d_r$ only once, either we will find the maximum x^* of p in the manifold or p will be determined as unbounded above. Furthermore, the order of the search is immaterial.

PROOF: From the above discussion any point in the r-dimensional manifold can be represented by $x' + \sum_{i=1}^{r} \gamma_i d_i$ for appropriate γ_i. Therefore, we must select the γ_i to maximize

$$p\left(x' + \sum_{i=1}^{r} \gamma_i d_i\right) = p(x') + \sum_{i=1}^{r} \{\tfrac{1}{2}\gamma_i^2 d_i^t Q d_i + \gamma_i d_i^t(Qx' + q)\}$$

Observe there are no terms $\gamma_i \gamma_j$, $i \neq j$, because the directions d_i are conjugate. Hence, the effect of searching in the positive and negative direction d_i is to find γ_i that maximizes

$$\{\tfrac{1}{2}\gamma_i^2 d_i^t Q d_i + \gamma_i d_i^t(Qx' + q)\}$$

Note also that the resultant value of γ_i is independent of the other terms in the function. Consequently, searching in each of the directions once only will either find the maximum or reveal p to be unbounded above. Also the order of the directions searched is immaterial. ◆

Conjugate Directions Maximize a Quadratic

A key implication of the above theorem is found by letting $r = n$. Then, because any n-dimensional manifold is E^n, the maximum of p over E^n is obtained in at most n searches. That n conjugate directions exist is shown in exercise 6.2. Also we will indicate later how to construct n conjugate directions. Thus conjugate directions can maximize a quadratic in at most n steps. Here, of course, by maximizing a quadratic we mean either calculating the optimal point or determining that p is unbounded above.

Algorithm for Maximizing a Quadratic

In terms of the previously developed notation, we may write the entire algorithm for maximizing p over E^n as

$$A_i = M^1 D_i \qquad i = 1, \ldots, n \tag{6.7}$$

where $D_i(x) = (x, d_i)$, $d_1, \ldots, d_n$, are conjugate, and $J = (-\infty, +\infty)$.† Having J of this form ensures the search is both in the positive and negative directions. Given a point x^1 for $i = 1, \ldots, n$ the algorithm calculates recursively

$$x^{i+1} \in A_i(x^i)$$

The point x^{n+1} will maximize p over E^n. Furthermore, it is easily seen that if p is unbounded above, the algorithm will reveal this, because, in such a case, for some map M^1 no point will maximize p (see exercise 6.9).

We now turn to developing procedures for calculating conjugate directions.

6.3. A CONJUGATE METHOD USING ONLY FUNCTION VALUES

This method determines conjugate directions for arbitrary quadratics without calculating derivatives and will be useful in exploiting quadratic approximations for general functions. To develop the theorem on which this

---

†The map M^1 was defined in Chap. 5.

method is based note that a given direction d may be used to generate different manifolds. As an example, suppose the directions d, d_1, and d_2 are linearly independent. Also suppose $x' \neq x''$ and neither x' nor x'' are in the subspace generated by d, d_1, and d_2. Then each of the following manifolds is different: the manifolds generated by x' and d; by x'' and d; by x', d, and d_1; by x'', d, and d_1; by x'', d, d_1, and d_2; etc. In general, whenever a direction d plus perhaps other directions is used to generate a given manifold we say the manifold *contains d*. Clearly, d can be contained in many different manifolds.

Theorem 6.2: *Let x' be a maximum of p in a manifold containing the direction d, and let x'' also be a maximum of p in a manifold containing d. Suppose*

$$p(x'') > p(x') \tag{6.8}$$

Then

$$d^1 = x'' - x'$$

is conjugate to d.

PROOF: First observe by (6.8) that x'' is not in the first manifold, hence d^1 and d must be linearly independent.

From definition of x'

$$\frac{dp}{d\tau}(x' + \tau d) = 0 \quad \text{at} \quad \tau = 0$$

and when this is written out

$$\tau d^t Q d + d^t(Qx' + q) = 0 \quad \text{at} \quad \tau = 0$$

Similarly for x''

$$\tau d^t Q d + d^t(Qx'' + q) = 0 \quad \text{at} \quad \tau = 0$$

Subtraction yields

$$d^t Q(x'' - x') = 0 \quad \blacklozenge$$

Implications of Theorem 6.2

The impact of this theorem can be described briefly. Let $d_1, \ldots, d_r$, $r < n$, be conjugate. Suppose both x' and x'' maximize p in manifolds containing $d_1, \ldots, d_r$ and $p(x') < p(x'')$. Then the directions

$$d_1, \ldots, d_r, d_{r+1}$$

where

$$d_{r+1} = x'' - x'$$

are conjugate. Thus, given r conjugate directions $r < n$, another conjugate direction is generated.

How might the points x' and x'' be obtained? Let x^1 be an arbitrary initial point. From this point suppose we determine the point x' after searching in the directions $d_1, \ldots, d_r$. Thus, from x^1 the point x' is obtained by the algorithm $x^{i+1} \in A_i(x^i)$, $i = 1, \ldots, r$, where A_i is defined in (6.7) and x' is set equal to x^{r+1}. Then by Theorem 6.1 x' will maximize p in the r-dimensional manifold generated by x^1 and the $d_1, \ldots, d_r$. If x' is not optimal, i.e., if x' does not maximize p over E^n, then a point x^2 may be determined such that

$$p(x') < p(x^2)$$

Starting from the point x^2 we again search in the directions $d_1, \ldots, d_r$ obtaining x''. The point x'' then maximizes p in a manifold generated by x^2 and $d_1, \ldots, d_r$.

By the monotonicity of the search,

$$p(x^2) \le p(x'')$$

Thus

$$p(x') < p(x'')$$

As both x' and x'' maximize p in manifolds containing $d_1, \ldots, d_r$, letting $d_{r+1} = x'' - x'$ the directions $d_1, \ldots, d_r, d_{r+1}$ are conjugate. Conse-

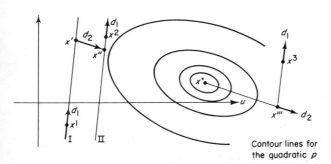

Fig. 6.1.

Conjugate directions maximize a quadratic.

The direction d_1 is indicated. Line I is the manifold generated by x^1 and d_1; while line II is the manifold generated by x^2 and d_1. The point x' maximizes p in manifold I, and x'' maximizes p in manifold II. Observe at x^2 to determine x'' required a search in the negative direction $-d_1$. The direction $d_2 = x'' - x'$.

To maximize p from any point, say x^3, maximize first in the positive and negative direction d_1, and at the resulting point, x''', do the same for d_2. This process determines x^*.

quently, we have a method for generating an additional conjugate direction (see Fig. 6.1).

Spacer Steps

In the procedure to follow it is important to have a step, such as that used above which, given a nonoptimal point x, generates a point y for which

$$p(y) > p(x)$$

We call this step a spacer step. The spacer step must also possess one other property. Should x be a solution, then the spacer step must indicate this fact. Thus the spacer step must first determine whether x is optimal, and if x is not optimal, it must calculate a better point. A typical example of a spacer step is one iteration of the cyclic-coordinate-ascent method, as choice of the cyclic method avoids calculation of derivatives.

A Conjugate-Direction Algorithm

We now state an algorithm that maximizes p in a finite number of iterations, where by maximizing p is meant that either p will be revealed to be unbounded or the optimal point found. Essentially, the algorithm simply repeats the above method for generating an additional conjugate direction n times. In the procedure the map $M^1(x, d)$ will use $J = (-\infty, +\infty)$. Also x_i^k and d_i^k are points in E^n.

A Procedure for Maximizing a Quadratic Function: 1

Initialization Step. Let an initial point x_0^0 and arbitrary direction $d_1 \neq 0$ be given. Calculate $x_0^1 \in M^1(x_0^0, d_1)$. Set $k = 1$.

Iteration k: x_0^k and $d_1, \ldots, d_k$ are given.
(a) Spacer step on x_0^k yielding x_1^k.
(b) For $i = 1, \ldots, k$ calculate recursively

$$x_{i+1}^k \in M^1(x_i^k, d_i)$$

(c) Set

$$d_{k+1} = x_{k+1}^k - x_0^k$$

(d) Determine

$$x_0^{k+1} \in M^1(x_{k+1}^k, d_{k+1})$$

Go to step (a) with $k + 1$ replacing k. Stop the procedure whenever the map M^1 indicates that p is unbounded above or the spacer step on x_0^k determines that x_0^k is a solution.

To interpret the procedure fix k. Observe that the point x_0^k is calculated after maximizing successively with $d_1, \ldots, d_k$. On iteration k the point x_{k+1}^k is also determined by maximizing successively with $d_1, \ldots, d_k$. The spacer step ensures that $p(x_0^k) < p(x_1^k)$ and by the monotonicity of the procedure $p(x_0^k) < p(x_{k+1}^k)$. Then from the reasoning given above, $d_{k+1} = x_{k+1}^k - x_0^k$ is conjugate to $d_1, \ldots, d_k$. Thus this procedure generates one more conjugate direction each iteration. At the end of the $(n - 1)$th iteration, n conjugate directions have been determined, and the procedure then must stop on step (a) of iteration n or before.

A Procedure for Maximizing a Quadratic Function: 2

Another procedure will now be stated that, as will be discussed subsequently, is easily adapted to maximizing an arbitrary f instead of just a quadratic. It will be stated first in terms of the quadratic p.

Procedure 2 differs from the previous procedure because n directions are always used. Initially these start out being n arbitrary directions $d_1^1, d_2^1, \ldots, d_n^1$. At the end of iteration one, the direction d_1^1 is deleted, the directions d_i^2, $i = 1, \ldots, n - 1$, are determined by $d_{i-1}^2 = d_i^1$ for $i = 2, \ldots, n$, and by making d_n^2 a new direction conjugate to d_{n-1}^2. The procedure continues in this manner. Again the map M^1 uses $J = (-\infty, +\infty)$.

Initialization Step. An initial point x_1^0 and n directions d_i^1, $i = 1, \ldots, n$, are given with $d_n^1 \neq 0$. For $i = 1, \ldots, n$ calculate $x_{i+1}^0 \in M^1(x_i^0, d_i^0)$. Define $x_{n+2}^0 = x_{n+1}^0$, and set $k = 1$.

Iteration k: x_{n+2}^{k-1} and d_i^k, $i = 1, \ldots, n$, are given.
(a) Spacer step on x_{n+2}^{k-1} yielding x_1^k.
(b) For $i = 1, \ldots, n$ calculate

$$x_{i+1}^k \in M^1(x_i^k, d_i^k)$$

(c) Set

$$d_{n+1}^k = x_{n+1}^k - x_{n+2}^{k-1}$$

and calculate

$$x_{n+2}^k \in M^1(x_{n+1}^k, d_{n+1}^k)$$

(d) Set $d_i^{k+1} = d_{i+1}^k$, $i = 1, \ldots, n$.
Go to iteration k with $k + 1$ replacing k.

Stop the procedure whenever the map M^1 indicates p to be unbounded above or the spacer step on x_{n+2}^{k-1} determines that x_{n+2}^{k-1} is a solution.

Interpretation

To establish that n conjugate directions are actually calculated, assume at the end of iteration k that the directions

$$d_{n+1}^k, d_n^k, d_{n-1}^k, \ldots, d_{n-k+1}^k$$

are conjugate. This means that the point x_{n+2}^k is a maximum of p in a manifold generated by $d_{n+1}^k, d_n^k, \ldots, d_{n-k+1}^k$. On iteration $k + 1$, $d_n^{k+1} = d_{n+1}^k, \ldots,$ $d_{n-k}^{k+1} = d_{n-k+1}^k$. Then the point x_{n+1}^{k+1} is also a maximum of p in a manifold generated by $d_{n+1}^k, \ldots, d_{n-k+1}^k$. Since the spacer step guarantees $p(x_{n+1}^{k+1}) > p(x_{n+2}^k)$, the direction $d_{n+1}^{k+1} = x_{n+1}^{k+1} - x_{n+2}^k$ is conjugate to $d_n^{k+1}, \ldots, d_{n-k}^{k+1}$. Thus at each iteration, if the procedure does not stop, another conjugate direction is generated (see also exercise 6.5).

Application to a General Function

Consider now a general function f instead of the quadratic p. Procedure 2 attempts to calculate conjugate directions for f; however, if f is not quadratic, then such a calculation is impossible. Nevertheless, the procedure can be thought of as calculating conjugate directions to a quadratic that approximates f. The quadratic approximation notion is thus being utilized. Should f either be quadratic or become quadratic, then the procedure determines n conjugate directions and converges in a finite number of steps.

Use as a Mixed Algorithm

Procedure 2 can be employed as a mixed algorithm for maximizing an arbitrary function f. Select the map of any one of the convergent algorithms in the previous chapter to be used for the spacer step. That algorithmic map will serve as the basic map B. Observe that as the iteration k approaches infinity, the map B, which is actually the spacer step, is used infinitely often. Convergence Theorem B then ensures that Procedure 2 converges. Consequently Procedure 2, by using the map of a convergent algorithm for the spacer step, is itself a convergent mixed algorithm for problem (5.1).

It should be observed that Procedure 2 does not require the calculation of any derivatives. Consequently, if the cyclic-coordinate-ascent method is used as the spacer step, no derivatives are used.

An Example

Procedure 2 will be illustrated by solving the following problem: where $x = (u, v)$

$$\min \quad p(u, v) = u - v + 2u^2 + 2uv + v^2$$

Observe the minimum instead of maximum. Letting $p(x) = q^t x + \frac{1}{2} x^t Q x$, we find

$$q = \begin{pmatrix} 1 \\ -1 \end{pmatrix} \quad \text{and} \quad Q = \begin{pmatrix} 4 & 2 \\ 2 & 2 \end{pmatrix}$$

Initialization Step. Let

$$x_1^0 = \begin{pmatrix} 0 \\ 0 \end{pmatrix} \quad d_1^1 = \begin{pmatrix} 0 \\ 0 \end{pmatrix} \quad \text{and} \quad d_2^1 = \begin{pmatrix} 1 \\ 0 \end{pmatrix}$$

Clearly, $x_2^0 = x_1^0$. To determine x_3^0, calculate

$$\min_{-\infty \leq \tau \leq +\infty} p(x_2^0 + \tau d_2^0) = \min \left\{ \tau(1, -1) \begin{pmatrix} 1 \\ 0 \end{pmatrix} + \frac{1}{2} \tau^2 (1, 0) \begin{pmatrix} 4 & 2 \\ 2 & 2 \end{pmatrix} \begin{pmatrix} 1 \\ 0 \end{pmatrix} \right\}$$

$$= \min \quad \{\tau + \frac{1}{2}\tau^2 4\}$$

and the optimal $\tau = -\frac{1}{4}$. Therefore,

$$x_3^0 = \begin{pmatrix} 0 \\ 0 \end{pmatrix} - \frac{1}{4} \begin{pmatrix} 1 \\ 0 \end{pmatrix} = \begin{pmatrix} -\frac{1}{4} \\ 0 \end{pmatrix}$$

We define $x_4^0 = x_3^0$.

Iteration 1.

$$x_4^0 = \begin{pmatrix} -\frac{1}{4} \\ 0 \end{pmatrix} \quad d_1^1 = \begin{pmatrix} 0 \\ 0 \end{pmatrix} \quad \text{and} \quad d_2^1 = \begin{pmatrix} 1 \\ 0 \end{pmatrix}$$

(a) Clearly x_4^0 is not a solution. To determine a better point let $d = \begin{pmatrix} 0 \\ 1 \end{pmatrix}$, then calculate x_1^1 to

$$\min_{-\infty \leq \tau \leq +\infty} p(x_4^0 + \tau d) = \min \left\{ (1, -1) \left[\begin{pmatrix} -\frac{1}{4} \\ 0 \end{pmatrix} + \tau \begin{pmatrix} 0 \\ 1 \end{pmatrix} \right] \right.$$

$$\left. + \frac{1}{2} \left[(-\frac{1}{4}, 0) + \tau(0, 1) \right] \begin{pmatrix} 4 & 2 \\ 2 & 2 \end{pmatrix} \left[\begin{pmatrix} -\frac{1}{4} \\ 0 \end{pmatrix} + \tau \begin{pmatrix} 0 \\ 1 \end{pmatrix} \right] \right\}$$

The optimal $\tau = \frac{3}{4}$. Therefore, $x_1^1 = x_4^0 + \frac{3}{4}d = \begin{pmatrix} -\frac{1}{4} \\ \frac{3}{4} \end{pmatrix}$

(b) Obviously $x_2^1 = x_1^1$. Also

$$x_3^1 = \begin{pmatrix} -\frac{5}{8} \\ \frac{3}{4} \end{pmatrix}$$

and

$$d_3^1 = x_3^1 - x_4^0 = \begin{pmatrix} -\frac{5}{8} \\ \frac{3}{4} \end{pmatrix} - \begin{pmatrix} -\frac{1}{4} \\ 0 \end{pmatrix} = \begin{pmatrix} -\frac{3}{8} \\ \frac{3}{4} \end{pmatrix}$$

We determine x_4^1 by

$$\min \quad p(x_3^1 + \tau d_3^1)$$

Then

$$x_4^1 = \begin{pmatrix} -1 \\ 1\frac{1}{2} \end{pmatrix}$$

(d)

$$d_1^2 = \begin{pmatrix} 1 \\ 0 \end{pmatrix} \qquad d_2^2 = \begin{pmatrix} -\frac{3}{8} \\ \frac{3}{4} \end{pmatrix}$$

Iteration 2. (a) It is easily seen that x_4^1 is a solution. Observe also that d_1^2 and d_2^2 are conjugate directions for Q.

This example illustrates that Procedure 2 optimizes a quadratic in a finite number of steps. No calculation of derivatives was required.

6.4. A CONJUGATE-GRADIENT METHOD

The previous procedure, when the spacer step is the cyclic-coordinate-ascent method, avoids calculation of derivatives. In situations for which derivatives are easily evaluated, however, the algorithm discussed in this section may be useful. Although a detailed discussion will be given shortly, the algorithm generates conjugate directions by use of the recursion

$$d_k = \nabla p(x^k) + \frac{\|\nabla p(x^k)\|^2}{\|\nabla p(x^{k-1})\|^2} d_{k-1}$$

in which $x^{k+1} \in M^1(x^k, d_k)$ and $d_1 = \nabla p(x^1)$. Also $\|\nabla p\|^2 = (\nabla p)^t \nabla p$.

Not only is the algorithm computationally valuable, but the theoretical development is quite interesting in its own right. Specifically, the theory requires discussion of both how to construct conjugate directions from linearly independent directions and how to construct orthogonal directions. The combination of these yields the algorithm. We assume that the matrix Q is negative definite, where by negative definite is meant $xQx < 0$ if $x \neq 0$.

Construction of Conjugate Directions from Linearly Independent Directions

Suppose we are given n linearly independent directions $e_1, \ldots, e_n$ and desire to form n Q-conjugate directions $d_1, \ldots, d_n$ from these linearly independent directions. A recursive procedure could be developed as follows:

Let $d_1 = e_1$ and for $k > 1$ define d_k in terms of e_k, the previously calculated d_i, $i = 1, \ldots, k-1$, and parameters $\gamma_{i,k}$ by

$$d_k = e_k + \sum_{i=1}^{k-1} \gamma_{i,k} d_i \tag{6.9}$$

The Q conjugacy requires $d_i^t Q d_k = 0$, $i = 1, \ldots, k-1$. Thus, because d_i, $i = 1, \ldots, k-1$, are already conjugate,

$$0 = d_i^t Q d_k = d_i^t Q e_k + \gamma_{i,k} d_i^t Q d_i \qquad i < k \tag{6.10}$$

and

$$\gamma_{i,k} = \frac{-d_i^t Q e_k}{d_i^t Q d_i}$$

Hence, the Q conjugate directions are recursively generated by the formula

$$d_k = e_k - \sum_{i=1}^{k-1} \left[\frac{d_i^t Q e_k}{d_i^t Q d_i} \right] d_i \tag{6.11}$$

For the application to be considered, the $d_i \neq 0$. Moreover, with Q negative definite, the denominator is not zero, and the d_i will be linearly independent (exercise 6.10).

Rewriting (6.11), we arrive at

$$e_k = d_k + \sum_{i=1}^{k-1} \left[\frac{d_i^t Q e_k}{d_i^t Q d_i} \right] d_i$$

and from the fact that the d_i are conjugate,

$$d_i^t Q e_k = 0 \quad \text{for} \quad i > k \tag{6.12}$$

Orthogonality Equivalent to I Conjugacy

If I is the identity matrix, then I conjugacy is identical with orthogonality. This follows because letting d_i^*, $i = 1, \ldots, n$, be I conjugate, we find

$$(d_i^*)^t I d_j^* = (d_i^*) d_j^* = 0 \qquad i \neq j$$

which is just the definition of orthogonality.

Thus using (6.10) and (6.11) with I replacing Q we obtain a method for generating orthogonal vectors.

Construction of Orthogonal Directions

Suppose given the $d_1, \ldots, d_n$ from (6.10) and (6.11) we wish to determine orthogonal vectors e_i', $i = 1, \ldots, n$, from the vectors $d_1, Qd_1, \ldots, Qd_{n-1}$ (observe the Q). Then set $e_1' = d_1$, and in general for $k > 1$, using (6.11),

$$e_k' = Qd_{k-1} - \sum_{i=1}^{k-1} \left[\frac{(e_i')^t Qd_{k-1}}{(e_i')^t e_i'} \right] e_i'$$

For our purposes a more convenient form is to define orthogonal vectors e_i^* recursively by $e_1^* = d_1$, and for $k > 1$

$$e_k^* = e_{k-1}^* - \frac{(e_{k-1}^*)^t e_{k-1}^*}{(e_{k-1}^*)^t Qd_{k-1}} \left[Qd_{k-1} - \sum_{i=1}^{k-2} \frac{(e_i^*)^t Qd_{k-1}}{(e_i^*)^t e_i^*} e_i^* \right] \tag{6.13}$$

Clearly e_i^* differs from e_i' by only a scale factor, and are themselves orthogonal. For the particular conjugate method to be developed, the term $(e_{k-1}^*)^t Qd_{k-1}$ will not be zero.

Just as Equation (6.12) was developed from (6.11)

$$(e_i^*)^t Qd_j = 0 \quad \text{for} \quad i > j + 1 \tag{6.14}$$

Simultaneous Generation of Orthogonal and Conjugate Directions

Now the linearly independent directions e_i and conjugate directions d_i will be chosen in a special manner to ensure that the e_i are themselves orthogonal. Initially let $e_1 = e_1^* = d_1$. Then using e_1 and Qd_1 obtain e_2^* from (6.13). Setting $e_2 = e_2^*$, certainly e_1 and e_2 are orthogonal.

Employing e_1 and d_1 determine d_2 from (6.11). Then d_1 and d_2 are conjugate. At this point e_1, e_2, and Qd_2 can be used to form e_3^* by (6.13). Set $e_3 = e_3^*$, then e_1, e_2, and e_3 are orthogonal.

In general $k - 1$ conjugate directions $d_1, \ldots, d_{k-1}$ and $k - 1$ orthogonal directions $e_1 = e_1^*, e_2 = e_2^*, \ldots, e_{k-1} = e_{k-1}^*$ are given. Via (6.13) generate $e_k = e_k^*$ from $e_1, e_2, \ldots, e_{k-1}$ and Qd_{k-1}. Then $e_1, \ldots, e_k$ are orthogonal. Now using (6.11) obtain d_k using $d_1, \ldots, d_{k-1}$ and e_k. Then the $d_1, \ldots, d_k$ are conjugate. Continuing this process, we obtain $d_i, i = 1, \ldots, n$, conjugate, and also the

$$e_i, i = 1, \ldots, n, \quad \text{are orthogonal} \tag{6.15}$$

Simplifications

Because the e_i are orthogonal, formulas (6.11) and (6.13) simplify greatly. Using (6.14), we find

$$e_k^t Q d_i = 0 \qquad k - 1 > i$$

so that (6.11) becomes

$$d_k = e_k - \frac{d_{k-1}^t Q e_k}{d_{k-1}^t Q d_{k-1}} d_{k-1} \tag{6.16}$$

Similarly from (6.12),

$$e_i^t Q d_{k-1} = 0 \quad \text{for} \quad i < k - 1$$

so that (6.13) reduces to

$$e_k = e_{k-1} - \gamma_{k-1} Q d_{k-1} \tag{6.17}$$

where

$$\gamma_{k-1} = \frac{e_{k-1}^t e_{k-1}}{e_{k-1}^t Q d_{k-1}}$$

Many important relationships exist between (6.16) and (6.17). It follows by the orthogonality of the e_i and from (6.17) that

$$e_k^t e_k = -\gamma_{k-1} e_k^t Q d_{k-1} \tag{6.18}$$

and

$$e_{k-1}^t e_{k-1} = \gamma_{k-1} e_{k-1}^t Q d_{k-1} \tag{6.19}$$

and using (6.16) and the conjugacy of the d_i

$$d_k^t Q d_k = d_k^t Q e_k \tag{6.20}$$

Equations (6.19) and (6.20) then yield

$$e_{k-1}^t e_{k-1} = \gamma_{k-1} d_{k-1}^t Q d_{k-1} \tag{6.21}$$

It will now be shown that

$$e_k^t d_{k-1} = 0 \tag{6.22}$$

As $e_1 = d_1$, from (6.17)

$$e_2 = d_1 - \frac{(d_1^t d_1) Q d_1}{d_1^t Q d_1}$$

so that

$$e_2^t d_1 = 0$$

Now assuming the induction hypothesis that $e_k^t d_{k-1} = 0$, when (6.16) is employed,

$$d_k^t e_k = e_k^t e_k \tag{6.23}$$

Then from (6.17)

$$e_{k+1}^t d_k = e_k^t d_k - \gamma_k d_k^t Q d_k$$

and via (6.21)

$$= 0$$

which establishes the induction. Consequently (6.22) is verified for all k. Equations (6.22) and (6.16) can now be combined to form

$$d_k^t e_k = e_k^t e_k \tag{6.24}$$

for all k.

Basic Relations of the Algorithm

Using these relationships, we may state (6.16) and (6.17) in their final form.

When (6.18) and (6.21) are inserted into (6.16)

$$d_k = e_k + \frac{\| e_k \|^2}{\| e_{k-1} \|^2} d_{k-1} \tag{6.25}$$

and rewriting (6.17)

$$e_k = e_{k-1} - \gamma_{k-1} Q d_{k-1} \tag{6.26}$$

where from (6.24) and (6.20)

$$\gamma_k = \frac{e_k^t d_k}{d_k^t Q d_k} \tag{6.27}$$

Equations (6.25) and (6.26) provide the basis of our conjugate gradient procedure for maximizing p. Essentially, given $d_1 = e_1$, using (6.25) and (6.26) recursively, we find the d_i generated are conjugate while the e_i are orthogonal.

The Conjugate-Gradient Algorithm

In the following the map M^1 uses $J = (-\infty, +\infty)$.

Initialization Step: Choose x^1 arbitrarily, let $d_1 = \nabla p(x^1)$, and determine $x^2 \in M^1(x^1, d_1)$. Set $k = 2$.

Step k: x^k is given. Define

$$d_k = \nabla p(x^k) + \frac{\|\nabla p(x^k)\|^2}{\|\nabla p(x^{k-1})\|^2} d_{k-1} \tag{6.28}$$

Calculate $x^{k+1} \in M^1(x^k, d_k)$.
Go to step k with $k + 1$ replacing k.
Stop the procedure when $\nabla p(x^k) = 0$.

Proof of Convergence

It now must be demonstrated that the conjugate-gradient procedure behaves precisely the same as Equations (6.25) and (6.26). If this is so the d_i generated will be conjugate, and the procedure will maximize a quadratic in a finite number of steps.

Some observations are required first. Given an arbitrary point x and direction d,

$$\begin{aligned} \nabla p(x + \tau d) &= q + Q(x + \tau d) \\ &= \nabla p(x) + \tau Q d \end{aligned} \tag{6.29}$$

Suppose $y \in M^1(x, d)$ and $y = x + \tau^0 d$. Then, since $J = (-\infty, +\infty)$,

$$p(x + \tau^0 d) = \max \ \{p(x + \tau d) | -\infty < \tau < +\infty\}$$

At $y = x + \tau^0 d$

$$0 = \frac{dp(x + \tau^0 d)}{d\tau} = \nabla p(x + \tau^0 d)^t d$$

and from (6.29)

$$0 = \nabla p(x + \tau^0 d)^t d = \nabla p(x)^t d + \tau^0 (Qd)^t d$$

Solving for τ^0, we get

$$\tau^0 = -\frac{d^t \nabla p(x)}{d^t Qd} \tag{6.30}$$

Therefore, the maximum of a quadratic q in the positive and negative direction d from x is found at $y = x + \tau^0 d$ for τ^0 in (6.30).

We now verify that the procedure generates directions d_i as specified by (6.25) and (6.26). If

$$e_k = \nabla p(x^k) \tag{6.31}$$

then Equation (6.25) holds via (6.28). It is therefore only necessary to prove that the procedure requires (6.26) to generate e_k as in Equation (6.31). Certainly $e_1 = d_1 = \nabla p(x^1)$. Assume the induction hypothesis that $e_i = \nabla p(x^i)$, $i \leq k$.

As the procedure calculates $x^{k+1} \in M^1(x^k, d_k)$, from (6.30)

$$x^{k+1} = x^k - \frac{d_k^t \nabla p(x^k)}{d_k^t Qd_k} d_k$$

by induction

$$= x^k - \frac{d_k^t e_k}{d_k^t Qd_k} d_k$$

and by (6.27)

$$= x^k - \gamma_k d_k$$

Then from (6.29)

$$\nabla p(x^{k+1}) = \nabla p(x^k) - \gamma_k Qd_k$$
$$= e_k - \gamma_k Qd_k$$

and Equation (6.31) is verified because when (6.26) is examined

$$e_{k+1} = \nabla p(x^{k+1})$$

The conjugate-gradient procedure is thus precisely the same as the recursion indicated by (6.25) and (6.26). Obviously the conjugate-gradient procedure is generating directions $d_i \neq 0$. This verifies that the directions in (6.11) and (6.28) cannot be zero. The e_i are also well defined.

Use as a Mixed Algorithm

Incorporating this conjugate-gradient technique into a mixed algorithm for maximizing an arbitrary f requires that it be coupled with a basic algorithm. The Cauchy procedure is appropriate because it requires partial derivatives as does the conjugate-gradient procedure. The two methods combined overcome the failure of the Cauchy method to provide convergence in a finite number of iterations for a quadratic. Moreover, from Convergence Theorem B by using the Cauchy procedure periodically the mixed algorithm will, in the limit, converge to a point at which f is at least as large as a solution point in the Cauchy procedure. Moreover, because of the conjugate directions, the convergence rate should also be enhanced over the Cauchy procedure alone.

EXERCISES

6.1. Complete the portion of the proof of Convergence Theorem B that refers to Convergence Theorem A.

6.2. Let Q be an $n \times n$ symmetric matrix with n eigenvectors e_i, $i = 1, \ldots, n$. Prove that the eigenvectors are conjugate directions.

6.3. Given a direction $d \neq 0$, let $y = x + \tau^0 d$ be such that

$$f(y) = \max_{\tau \in J} \ \{f(x + \tau d)\}$$

where $J = [-\alpha, +\alpha]$ and $\alpha > |\tau^0|$. Show that

$$\nabla f(y)^t d = 0$$

What could occur if
    (a) $J = [0, \alpha]$
    (b) $d = 0$
    (c) $J = [-\beta, +\beta]$ where $0 \leq \beta < |\tau^0|$

6.4. Maximize the function $p(x) = q^t x - \frac{1}{2} x^t Q x$ where

$$q = \begin{bmatrix} 1 \\ 3 \\ 5 \end{bmatrix} \qquad Q = \begin{bmatrix} 10 & 6 & 1 \\ 6 & 4 & 0 \\ 1 & 0 & 1 \end{bmatrix}$$

with

    (a) The modified Newton method.

    (b) Each of the conjugate methods described in the text.

    (c) Conjugate directions constructed from the 3-unit coordinate vectors

$$\begin{bmatrix} 1 \\ 0 \\ 0 \end{bmatrix} \begin{bmatrix} 0 \\ 1 \\ 0 \end{bmatrix} \quad \text{and} \quad \begin{bmatrix} 0 \\ 0 \\ 1 \end{bmatrix}$$

6.5. Prove that the procedure for maximizing a quadratic function: 2 (Sec. 6.3) becomes the procedure for maximizing a quadratic function: 1 if $d_i^1 = 0$, $i = 1, \ldots, n - 1$.

6.6. *Diagonalizing a Matrix.* Let Q be a positive definite $n \times n$ matrix. Show how to determine an $n \times n$ matrix E such that $E^t Q E$ is diagonal. Use conjugate directions.

6.7. Let Q be a positive definite matrix. Discover a procedure based on conjugate directions for solving the equations

$$Qx = b$$

6.8. *Construction of a Positive Definite or Semidefinite Matrix.* Let B be an $m \times n$, $m \geq n$, matrix with linearly independent columns. Prove the $n \times n$ matrix

$$Q = B^t B$$

is symmetric and positive definite.

    Hint: $x^t(B^t B)x = \| Bx \|^2$, and $Bx = 0$ only if $x = 0$ by independence.

    Construct a 3×3 negative definite matrix. How could you construct a positive semidefinite matrix?

6.9. For p quadratic consider the problem of finding $y \in M^1(x, d)$ for $J = (-\infty, +\infty)$. This problem can be written

$$\max_{-\infty < \tau < +\infty} p(x + \tau d)$$

Show

$$p(x + \tau d) = a\tau^2 + b\tau + c$$

where a, b, and c are constants independent of τ. For what values of a is p unbounded from above on the line $x + \tau d$? For what values of a can the maximum be obtained by differentiation? What happens if $b = 0$?

6.10. Suppose Q is negative definite. Let d_i, $i = 1, \ldots, r$, be such that $d_i \neq 0$, and $d_i Q d_j = 0$ if $i \neq j$. Prove the d_i are conjugate directions.

6.11. Does the conjugate-gradient method converge for Q not negative definite?

NOTES AND REFERENCES

§6.1. Convergence Theorem B is a straightforward extension of Convergence Theorem A.

§6.2. This presentation differs from previous discussions of conjugate directions in that Q is not restricted to be negative definite. Theorem 6.1 is based upon a modification found in ZANGWILL (1967f) of the theorem in POWELL (1964). Theorem 6.2 is also a modification of a theorem in POWELL (1964).

§6.3. This section is adapted from ZANGWILL (1967f) [see also POWELL (1964)].

§6.4. BECKMAN developed this procedure in his article. For some computational results, refer to FLETCHER and REEVES.

ADDITIONAL COMMENTS

Conjugate directions were suggested in 1952 to solve linear equations in HESTENES and STIEFEL (see exercise 6.7). Other conjugate-direction techniques include those in ROSENBROCK and a paper related to that of Rosenbrock by FLETCHER and POWELL [see also FLETCHER (1964); POWELL (1962); STIEFEL; MARTIN and TEE and for a review FLETCHER (1965)].

Other approaches include those of COLLATZ; NEDLER and MEAD; HIMSWORTH, *et al.* and the partan method by SHAH, *et al.* (1961), (1964). A comparison of unconstrained methods can be found in BROOKS.

Information about manifolds can be obtained from standard linear-algebra texts.

7 Continuity, Compactness, and Closedness

This brief chapter presents several mathematical results useful in the remainder of the text. We study the supremum and infimum operations and indicate how they extend the maximum and minimum operations respectively. Certain theorems that state how continuity and closedness may be preserved through the maximum operation are also proved. These powerful theorems will be used repeatedly in the following chapters.

7.1. MAXIMA VERSUS SUPREMA

We have previously assumed when the function $h(w)$ is maximized over a set S that a point w^* in S exists such that

$$h(w^*) = \max \ \{h(w) \,|\, w \in S\}$$

For almost all practical cases, this assumption is true. However, to maintain mathematical rigor we must investigate when such a maximizing point does or does not exist. The concepts of continuity and compactness will turn out to be relevant.

First some examples for $w \in E^1$.

(a)   max $\{h(w) = w \mid 0 \leq w \leq 1\}$
(b)   max $\{h(w) = w \mid 0 \leq w < 1\}$
(c)   max $\{h(w) = w^2 \mid 0 \leq w < +\infty\}$
(d)   max $\{h(w) = 1 - e^{-w} \mid 0 < w < +\infty\}$

For (a) the point $w^* = 1$ maximizes. In (b) no point in the set $0 \leq w < 1$ achieves the maximum, as the point $w = 1$ is not in the set. Also in both (c) and (d) there is no real number w at which the maximum is achieved. Because $+\infty$ and $-\infty$ are not real numbers, they are, of course, excluded. To be precise, the mathematical operation of maximization is not defined for (b), (c), and (d). Instead, the concept of supremum (sup), which is a generalization of maximum, must be introduced.

Definition: Let $h: V \rightarrow E^1$ be a function. For a set $S \subset V$ define γ, where

$$\gamma = \sup \; \{h(w) \mid w \in S\}$$

to be the **supremum** of h over S if

(a) $\gamma \geq h(w)$ for all $w \in S$ and
(b) there is a sequence $w^k \in S$ such that

$$\lim_{k \to \infty} h(w^k) = \gamma$$

Discussion

The scalar γ may assume the values $+\infty$ or $-\infty$.
Suppose, as in case (a), there is actually a maximizing point w^*, then the supremum $\gamma = h(w^*)$, and the sequence w^k is defined by

$$w^k = w^* \quad \text{for all} \quad k$$

Thus, when a maximizing point $w^* \in S$ exists, $h(w^*)$ is the supremum as well as the maximum.
Cases (b), (c), and (d) may now be modified to
(b') sup $\{h(w) = w \,|\, 0 \leq w < 1\} = 1$
(c') sup $\{h(w) = w^2 \,|\, 0 \leq w < +\infty\} = +\infty$
(d') sup $\{h(w) = 1 - e^{-w} \,|\, 0 < w < +\infty\} = 1$
In case (b') the sequence w^k may be taken to be

$$w^k = 1 - \frac{1}{k} \tag{7.1}$$

while for (c') and (d') let

$$w^k = k \tag{7.2}$$

It is important to observe that in both (7.1) and (7.2)

$$\lim_{k \to \infty} w^k$$

is not in the set being considered. In (7.1)

$$\lim \quad w^k = 1$$

and 1 is not in the set $0 \leq w < 1$. Similarly in (7.2)

$$\lim \quad w^k = +\infty$$

which is not in the sets for (c') or (d').

7.2. CONTINUITY, COMPACTNESS, AND MAXIMIZATION

The following important theorem states conditions under which the maximum operation may be used instead of a supremum. Specifically, a continuous function on a compact set achieves its maximum.

Theorem 7.1: *Let* $h: V \rightarrow E^1$ *be a continuous function. Suppose*

$$\gamma = \sup \ \{h(w)\,|\,w \in S\}$$

where $S \subset V$ *and* S *is a compact set. Then there is a point* w^* *in* S *that maximizes* h *over* S.

PROOF: By definition of supremum there is a sequence $\{w^k\}_1^\infty$ in S such that

$$\lim_{k \rightarrow \infty} \ h(w^k) = \gamma \tag{7.3}$$

Via compactness of S a $\mathscr{K}$ exists such that

$$w^k \longrightarrow w^* \qquad k \in \mathscr{K}$$

where $w^* \in S$. Continuity of h then implies

$$\lim_{k \in \mathscr{K}} \ h(w^k) = h(w^*) \tag{7.4}$$

and as $\mathscr{K}$ is an infinite subset of the positive integers

$$\lim_{k \rightarrow \infty} \ h(w^k) = \lim_{k \in \mathscr{K}} \ h(w^k) \tag{7.5}$$

Summarizing (7.3), (7.4), and (7.5)

$$h(w^*) = \gamma \tag{7.6}$$

From the definition of supremum, for all $w \in S$

$$h(w) \leq \gamma$$

Consequently, as $w^* \in S$, w^* maximizes h over S. ◆

Infimum

Just as supremum extends the concept of maximum, the operation infimum (inf) deals with functions that do not achieve their minimum. More precisely, let

$$\inf \ \{h(w) \,|\, w \in S\} = -\sup \ \{-h(w) \,|\, w \in S\}$$

Corollary 7.1.1: *Let* $h: V \to E^1$ *be a continuous function and* $S \subset V$ *where* S *is a compact set. Suppose*

$$h(w) > 0 \quad \text{for all} \quad w \in S$$

Then

$$\inf \ \{h(w) \,|\, w \in S\} > 0$$

PROOF: (see exercise 7.3)

The relationships among continuity, compactness, and suprema are extended in the next theorem. But before stating the theorem, we must first define the Cartesian product of two sets: Given two sets U and V, the *Cartesian-product* set

$$W = U \otimes V$$

is the set of all points $w = (u, v)$ for $u \in U$ and $v \in V$. As an example, the set $E^2 = E^1 \otimes E^1$, and $E^{n+m} = E^n \otimes E^m$ (see exercises 7.1 and 7.2). A function $h(u, v): U \otimes V \to E^1$ will be defined for all points (u, v) for which $u \in U$ and $v \in V$.

Theorem 7.2: *Let* $h(u, v): U \otimes V \to E^1$ *be continuous on* $W = U \otimes V$. *Define a function* $j(u): U \to E^1$ *by*

$$j(u) = \sup \ \{h(u, v) \,|\, v \in V\} \tag{7.7}$$

If V *is compact then* j *is itself continuous on* U.

PROOF: Let $u^k \to u^\infty$, it must be shown that

$$\lim_{k \to \infty} \ j(u^k) = j(u^\infty) \tag{7.8}$$

Given an arbitrary subset $\mathcal{K}$, we will demonstrate that for some $\mathcal{K}^1 \subset \mathcal{K}$

$$\lim_{k \in \mathcal{K}^1} \ j(u^k) = j(u^\infty) \tag{7.9}$$

then, since $\mathcal{K}$ is arbitrary, (7.8) will be proved.

By compactness of V and Theorem 7.1 for some $v^k \in V$

$$j(u^k) = h(u^k, v^k) = \sup \ \{h(u^k, v) | v \in V\} \tag{7.10}$$

Again because V is compact there is a $\mathcal{K}^1 \subset \mathcal{K}$ such that

$$v^k \longrightarrow v^\infty \qquad k \in \mathcal{K}^1$$

and $v^\infty \in V$.
   By the definition of v^k,

$$h(u^k, v^k) \geq h(u^k, v) \quad \text{for all} \quad v \in V \tag{7.11}$$

and taking limits of (7.11), because h is continuous, we find

$$h(u^\infty, v^\infty) \geq h(u^\infty, v) \quad \text{for all} \quad v \in V$$

Hence v^∞ maximizes $h(u^\infty, v)$ over $v \in V$, and

$$j(u^\infty) = h(u^\infty, v^\infty) \tag{7.12}$$

Via continuity of h we may also take limits of (7.10) yielding

$$\lim_{k \in \mathcal{K}^1} j(u^k) = \lim_{k \in \mathcal{K}^1} h(u^k, v^k) = h(u^\infty, v^\infty) \tag{7.13}$$

Consequently from (7.12) and (7.13)

$$\lim_{k \in \mathcal{K}^1} j(u^k) = j(u^\infty) \tag{7.14}$$

and Equation (7.9) is verified. ◆

   Functions of the form in (7.7) arise frequently in NLP algorithms. For example, consider the map M^1 redefined as follows: Let

$$f: E^{2n} \otimes J \longrightarrow E^1$$

be

$$f(x + \tau d)$$

Here $(x, d) \in E^{2n}$, and $\tau \in J$. Then $M^1(x, d)$ is the set of all $y = x + \tau d$ for $\tau \in J$ such that

$$j(y) = \sup \ \{f(x + \tau d) | \tau \in J\} \tag{7.15}$$

Other examples of the function j are in exercise 7.4.
   Lemma 7.3 generalizes Lemma 5.1, and its proof is left as an exercise.

Lemma 7.3: *Suppose* $h: U \otimes V \to E^1$ *is a continuous function on* $U \otimes V$. *For V, a compact set, let*

$$j(u) = \max \ \{h(u, v) | v \in V\} \tag{7.16}$$

and define the point-to-set map $H: U \to V$ *by*

$$H(u) = \{v | j(u) = h(u, v), v \in V\} \tag{7.17}$$

Then the map H is closed on U.

Examples of H appear in exercises 7.4 and 7.5.

These results on continuity, compactness, and suprema will be of great value in the subsequent chapters.

EXERCISES

7.1. What is the Cartesian product $U \otimes V$, for $U \subset E^1$ and $V \subset E^1$ where
    (a) $U = [0, 1]$ and $V = [0, 1]$
    (b) $U = [0, 1] \cup [2, 3]$ and $V = [0, 1]$
    (c) $U = [0, +\infty)$ and $V = [0, +\infty)$

7.2. Find $U \otimes V$ where $U \subset E^1$ and $V \subset E^2$
    (a) $U = [0, 1] \cup [2, 3]$ $V = \{(x, y) | x^2 + y^2 \leq 1, x, y \in E^1\}$
    (b) $U = [0, 1]$ $V = [0, 1] \otimes [0, 1]$

7.3. Prove Corollary 7.1.1 and Lemma 7.3.

7.4. Let h be continuous, $h(u, v): U \otimes V \to E^1$. Define, as in the text,

$$j(u) = \ \sup \ \{h(u, v) | v \in V\}$$

and

$$H(u) = \{v | j(u) = h(u, v)\}$$

where $j: U \to E^1$ and H is a point-to-set map $U \to V$. Calculate j and H when
    (a) $U = [0, 1]$ $V = [0, 1]$ $h = u^2 + v^2$
    (b) $U = [-1, 1]$ $V = [0, +\infty)$ $h = uv - v^2$
    (c) $U = [-1, +1]$ $V = [-1, +1]$ $h = uv$
    (d) $U = [-1, +1]$ $V = (-\infty, +\infty)$ $h = uv$

7.5. Let an objective function f be continuously differentiable, and suppose the feasible region F is compact. Define the function

$$\nabla f(x)^t(y - x) \quad \text{on} \quad F \otimes F$$

Here: $x \in F$ and $y \in F$.
Let

$$j(x) = \sup \ \{\nabla f(x)^t(y - x) \,|\, y \in F\}$$

and

$$H(x) = \{y \,|\, j(x) = \nabla f(x)^t(y - x), y \in F\}$$

Prove j is continuous and H is a closed map.

NOTES AND REFERENCES

The topics in this chapter may be found in standard texts such as APOSTLE; BERGE; SIMMONS. The treatment by Berge is especially relevant.

8   Some Procedures

for

Linear Constraints

The nonlinear programming (NLP) problem with linear constraints may be stated as

$$\max \quad f(x)$$
$$\text{subject to} \quad Ax = b \tag{8.1}$$
$$x \geq 0$$

where A is an $m \times n$ matrix, x is an n vector, and b is an m vector. Of the various NLP problems, problems with only linear constraints arise frequently and are generally easier to solve than problems with nonlinear constraints.

In this chapter we present several methods for solving problem (8.1). The first is based upon linear approximations. The second, the convex-simplex method, extends the linear-simplex method to nonlinear objective functions. Finally, we discuss the manifold-suboptimization method, which by considering only some of the constraints at each stage, solves problem (8.1) through the solution of smaller subproblems.

8.1. A LINEAR-APPROXIMATION METHOD

Perhaps the most straightforward approach to solving problem (8.1) is by linear approximations. Given a feasible point x^k, let y^k be the solution to the linear programming (LP) problem that uses the same constraints as problem (8.1) but whose objective function is a linear approximation to f at x^k. Then the direction $d^k = y^k - x^k$ is a good direction to seek an increased value of f.

At any feasible point x^k the linear approximation to $f(y)$ is $f_L(y)$ where

$$f_L(y) = f(x^k) + \nabla f(x^k)^t(y - x^k)$$

The algorithm, given f_L and x^k fixed, obtains the solution y^k to the sub-problem

$$\max \quad f(x^k) + \nabla f(x^k)^t(y - x^k)$$
$$\text{subject to} \quad Ay = b \tag{8.2}$$
$$y \geq 0$$

But since several of the terms in the objective function are constant, the algorithm actually solves the simpler but equivalent LP subproblem

$$\max \quad \nabla f(x)^t y$$
$$\text{subject to} \quad Ay = b \tag{8.3}$$
$$y \geq 0$$

where x is a fixed feasible point and y is the variable.

Then using the direction

$$d^k = y^k - x^k$$

we arrive at the next point $x^{k+1} \in M^1(x^k, d^k)$. Here $J = [0, 1]$. Observe that both x^k and y^k are feasible. Therefore, with $J = [0, 1]$, any point

$$x^k + \tau(y^k - x^k) \qquad \tau \in J \tag{8.4}$$

is feasible, so that x^{k+1} is also feasible. We now state the algorithm precisely.

The Algorithmic Map

The algorithm is

$$A = M^1 D$$

where $J = [0, 1]$, and the map $D(x^k) = (x^k, d^k)$ is obtained by letting

$$d^k = y^k - x^k \tag{8.5}$$

where y^k solves (8.3) for a feasible point $x = x^k$.

We call a feasible point x^* a solution if y^*, a solution to subproblem (8.3) with $x = x^*$, implies

$$\nabla f(x^*)'(y^* - x^*) \leq 0 \tag{8.6}$$

It is left as an exercise to prove that the Kuhn-Tucker (K-T) conditions hold at a solution (see exercise 8.1).

Should x^k be a solution, terminate the algorithm. Also, to begin the algorithm, an initial feasible point x^1 is given.

Convergence Proof

To prove convergence assume that f is continuously differentiable. Assuming that the feasible region is compact ensures that condition 1 of Convergence Theorem A holds, because given x^1 feasible, all points x^k must be feasible.

Condition 2

Let $z = x$ and $Z(z) = f(x)$. Clearly condition 2(b) holds.

To prove 2(a) assume x' is not a solution. Then letting y' solve (8.3) for $x = x'$

$$\nabla f(x')^t(y' - x') > 0 \qquad (8.7)$$

but because $d' = y' - x'$, if $w \in M^1(x', d')$,

$$f(w) > f(x') \qquad (8.8)$$

Condition 3

Finally the map A must be proved closed. First D will be studied. Let

$$x^k \longrightarrow x^\infty \qquad k \in \mathcal{K}$$

and

$$d^k \longrightarrow d^\infty \qquad k \in \mathcal{K}$$

Then

$$y^k \longrightarrow y^\infty = d^\infty - x^\infty \qquad k \in \mathcal{K}$$

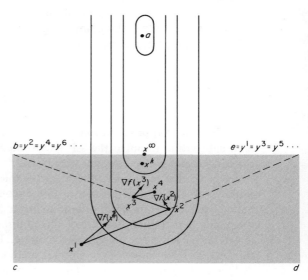

Fig. 8.1.

The linear approximation algorithm. The point a gives the unconstrained maximum of f, while the ovals are f's contour lines. The feasible region is the rectangle b, c, d, e. A detailed description of the generation of x^1, x^2, and x^3 is given. Observe that the sequence x^k converges to x^∞, the solution point.

To prove D is closed it is sufficient to establish that y^∞ solves problem (8.3) where $x = x^\infty$. Then, because $d^\infty = y^\infty - x^\infty$,

$$(x^\infty, d^\infty) \in D(x^\infty)$$

Since all y^k are feasible, y^∞ is also feasible. Moreover, by the definition of y^k,

$$\nabla f(x^k)^t (y^k - x^k) \geq \nabla f(x^k)^t (y - x^k) \tag{8.9}$$

for any feasible y. Taking limits in (8.9), as f is continuously differentiable, we arrive at

$$\nabla f(x^\infty)^t (y^\infty - x^\infty) \geq \nabla f(x^\infty)^t (y - x^\infty) \tag{8.10}$$

But then, because y^∞ is feasible and (8.10) holds for all feasible y, y^∞ solves the subproblem for $x = x^\infty$. Consequently, D is closed. (For an alternative proof of closedness see exercise 8.4.)

Since all x^k and y^k are feasible and thus on a compact set, all (x^k, d^k) are in a compact set. With M^1 and D closed Corollary 4.2.1 proves that A is closed. The procedure therefore satisfies Convergence Theorem A.

A graphic representation of this algorithm's behavior is depicted in Fig. 8.1. Observe that the points x^k generated zig-zag back and forth as the solution point x^∞ is reached. Such zig-zagging often retards convergence.

8.2. The Convex-Simplex Method

The linear-approximation method utilizes the linear-simplex method as a subroutine to solve the LP subproblem. The convex-simplex method instead attempts to operate within the linear-simplex-method structure. Indeed, it behaves just like the linear-simplex method whenever any linear portions of the objective function are encountered. The method was originally posed for the problem of minimizing a convex function subject to linear inequality constraints. Because of this and its linear-simplex-method nature, it was termed the convex-simplex method. For consistency with the rest of this text, however, the convex-simplex method will be described for problem (8.1) with f a general function having continuous partial derivatives.†

---

†It may be noted that in the next chapter we will discuss how to add conjugate direction moves to the convex-simplex method. With conjugate directions the convex-simplex method uses quadratic approximations for general f and converges in a finite number of steps if f is quadratic.

Rationale of the Convex-Simplex Method

The convex-simplex method will now be motivated by an informal discussion. Following this heuristic explanation, we will present an exact mathematical statement of the method. Our intention is to follow the linear-simplex method for problem (8.1) as far as it will go and to modify it only as the nonlinearity of the objective function requires. Generally, subscripting a vector indicates a component of the vector, e.g., b_i is the ith component of b.

It is assumed without loss of generality that a standard LP phase I procedure has insured that all rows of the matrix A are linearly independent and has generated an initial basic feasible solution x^1, a tableau T^1, and a corresponding right-hand-side b^1 such that

$$T^1x^1 = b^1$$
$$x^1 \geq 0 \tag{8.11}$$

The convex-simplex method will generate a sequence of tableaux each new one obtained by standard pivot techniques from the previous one. It may be assumed that each tableau T is an $m \times n$ matrix. Let B_j be the number of the column associated with the jth basic variable, $1 \leq j \leq m$, then the B_jth column is a column of all zeros except for a one in the jth place. Since x^1 is a basic solution,

$$x_{B_j}^1 = b_j^1 \qquad j \in \mu$$

where $x_{B_j}^1$ is a basic variable and μ is the set of the first m positive integers. All nonbasic variables in the vector x^1 are zero.

In the linear-simplex method the relative-cost vector corresponding to tableau T^1 would now be calculated and either x^1 declared optimal or a nonbasic variable increased. With a nonlinear objective function the relative-cost vector can be calculated using the appropriate gradient of f. For a given tableau T let

$$\nabla f(x)_B = \left(\frac{\partial f(x)}{\partial x_{B_1}}, \frac{\partial f(x)}{\partial x_{B_2}}, \ldots, \frac{\partial f(x)}{\partial x_{B_m}} \right)^t \tag{8.12}$$

Then the relative costs of x for the given tableau T are

$$c(x) = (\nabla f(x) - T^t \nabla f(x)_B) \tag{8.13}$$

where $\nabla f(x)$ is the gradient of f evaluated at x (see exercise 8.10). It will always be clear from the context with respect to which tableau the relative costs are taken. It is convenient on the kth iteration to let $c^k = c(x^k)$.

For tableau $T^1 = (t^1_{ij})$, we would obtain the relative cost vector c^1. In the linear-simplex method the value

$$c^1_s = \max \ \{c^1_i \,|\, i \in \nu\}$$

would be calculated, where ν is the set of first n positive integers. Should c^1_s be nonpositive, x^1 is, as will be shown later, a solution. Otherwise, we increase the variable x_s adjusting only basic variables, and the objective function increases. In the linear-simplex method since the objective function is linear, one is assured that the objective function will increase as long as x_s is increased. If $t^1_{si} \leq 0$ for all $i \in \mu$, then x_s may be indefinitely increased yielding an unbounded objective function. Otherwise x_s is increased until a basic variable, say x_{B_r}, becomes zero. The tableau is then transformed by pivoting on t^1_{rs}. Note that the components of x that are in the new basis are those components of largest magnitude. That is, the m largest components of x are in the basis.

With a nonlinear objective function if $c^1_s > 0$, then increasing x_s and adjusting only basic variables, while initially producing an increase in f, may, should x_s be increased too far, actually cause f to decrease. The convex-simplex method increases x_s either until further increase would no longer increase f or until a basic variable becomes zero. The value of x at which the first of these two events occurs is the point x^2.

The precise procedure is as follows. Either $t^1_{is} > 0$ for some i or not. If there exists an i such that $t^1_{is} > 0$, then increasing x_s must force a basic variable to zero. Let y^1 be the value of x obtained by increasing x_s until a basic variable becomes zero. The value x^2 is then obtained from the formula

$$f(x^2) = \max \ \{f(x^1 + \lambda(y^1 - x^1)) \,|\, 0 \leq \lambda \leq 1\} \tag{8.14}$$

On the other hand, it may be that $t^1_{is} \leq 0$, $i \in \mu$. Then x_s may be increased indefinitely without driving a basic variable to zero, and increasing x_s will geometrically form a ray emanating from x^1. In this case let $y^1 \neq x^1$ be any point on that ray. Determine a point x^2 on that ray that maximizes f. Specifically

$$f(x^2) = \max \ \{f(x^1 + \lambda(y^1 - x^1)) \,|\, \alpha \geq \lambda \geq 0\}$$

for α very large.

After x^2 is calculated, select as tableau T^2 any tableau with the largest (in magnitude) components of x forming the basis.

Iteration 1 is completed, and iteration 2 commences. Observe that if a basic variable was not driven to zero, the point x^2 with respect to tableau T^2 may have a positive nonbasic variable. That is, some nonbasic variable, say x^2_r, may be positive. For generality assume this is so. Again calculate the relative-cost vector

$$c^2 = (\nabla f(x^2) - (T^2)^t \nabla f(x^2)_B) \tag{8.15}$$

If the positive nonbasic variable has a negative relative-cost factor, then decreasing that variable and adjusting only basic variables will increase f. Similarly, if any nonbasic variable has a positive relative-cost factor, increasing it will increase f. Let

$$c_{s_1}^2 = \max \ \{c_i^2 \,|\, i \in v\} \tag{8.16}$$

and

$$c_{s_2}^2 \cdot x_{s_2}^2 = \min \ \{c_i^2 \cdot x_i^2 \,|\, i \in v\} \tag{8.17}$$

If $c_{s_1}^2 > 0$, then x_{s_1} is an excellent choice to increase. If $c_{s_2}^2 \cdot x_{s_2}^2 < 0$, then $x_{s_2}^2$ must be positive and x_{s_2} is an excellent choice to decrease.

We will give a specific rule for selecting $x_s = x_{s_1}$ or $x_s = x_{s_2}$ shortly. Should $x_s = x_{s_1}$, increase x_s until either f no longer increases or a basic variable becomes zero. Should $x_s = x_{s_2}$, decrease x_s until either f no longer increases, x_s itself becomes zero, or a basic variable becomes zero. Call the corresponding value of the x variable x^3. The procedure then continues in a manner analogous to iteration 2.

We now give a precise statement of the convex-simplex method.

The Algorithmic Procedure

In the procedure any tie may be arbitrarily broken.

Initialization Step. An appropriate linear-simplex method phase I procedure has generated a basic feasible solution x^1 with corresponding tableau T^1. Go to Step 1 of iteration k with $k = 1$.

Iteration k. The feasible point x^k and tableau $T^k = (t_{ij}^k)$ are given.

Step 1: Calculate the relative-cost vector: Let

$$c^k = (\nabla f(x^k) - (T^k)^t \nabla f(x^k)_B) \tag{8.18}$$

$$c_{s_1}^k = \max \ \{c_i^k \,|\, i \in v\}$$

and

$$c_{s_2}^k x_{s_2}^k = \min \ \{c_i^k x_i^k \,|\, i \in v\}$$

If $c_{s_1}^k = c_{s_2}^k x_{s_2}^k = 0$, terminate. Otherwise, go to step 2.

Step 2: Determine the nonbasic variable to change:
If $c_{s_1}^k \geq |c_{s_2}^k x_{s_2}^k|$, increase $x_s = x_{s_1}$ adjusting only basic variables.
If $c_{s_1}^k \leq |c_{s_2}^k x_{s_2}^k|$, decrease $x_s = x_{s_2}$ adjusting only basic variables.

Step 3: Calculate x^{k+1}: There are three cases to consider.

Case(a) x_s is to be increased, and for some i, $t_{is}^k > 0$.

Increasing x_s will drive a basic variable to zero. Let y^k be the x value when that occurs. Specifically,

$$y_i^k = x_i^k \qquad i \in N - s \tag{8.19}$$

$$y_s^k = x_s^k + \Delta^k \tag{8.19a}$$

$$y_{B_i}^k = x_{B_i}^k - t_{is}^k \Delta^k \qquad i \in \mu \tag{8.19b}$$

where *N-s* is the set of indices of the nonbasic variables except s and

$$\Delta^k = \frac{x_{B_r}^k}{t_{rs}^k} = \min \left\{ \frac{x_{B_i}^k}{t_{is}^k} \Big| t_{is}^k > 0 \right\} \tag{8.20}$$

Find

$$x^{k+1} \in M^1(x^k, d^k) \tag{8.21}$$

where $J = [0, 1]$ and

$$d^k = y^k - x^k$$

Case (b) x_s is to be increased, and $t_{is}^k \leq 0$ for all i.

In this case x_s may be increased indefinitely without driving a basic variable to zero. Define y^k as in Equation (8.19) except let $\Delta^k = 1$. Determine x^{k+1} such that

$$x^{k+1} \in M^1(x^k, d^k)$$

where $J = [0, a]$ for a very large and

$$d^k = y^k - x^k$$

Case (c) x_s is decreased.

Determine y^k using Equation (8.19) except define Δ^k as follows.

$$\Delta^k = \max \{\Delta_1^k, \Delta_2^k\}$$

where

$$\Delta_1^k = \frac{x_{B_r}^k}{t_{rs}^k} = \max \left\{ \frac{x_{B_i}^k}{t_{is}^k} \Big| t_{is}^k < 0 \right\}$$

and

$$\Delta_2^k = -x_s^k$$

Should $t_{is}^k \geq 0$, $i \in \mu$, let $\Delta_1^k = -\infty$. Here y^k is the x corresponding to the point where, as x_s is decreased, either a basic variable becomes zero or x_s itself becomes zero, whichever occurs first. Calculate x^{k+1} using (8.21).

Step 4: After x^{k+1} is calculated, determine tableau T^{k+1} as indicated in the next paragraph, and go to iteration k with $k + 1$ replacing k.

Tableau Selection

In any of the above cases the tableau T^{k+1} is determined from x^{k+1} in the following manner. Order the components of x^{k+1} by magnitude

$$x_{i_1}^{k+1} \geq x_{i_2}^{k+1} \geq \cdots \geq x_{i_n}^{k+1}$$

Determine the m largest in magnitude. Let T^{k+1} be any tableau such that these m are in the basis for that tableau. Thus $x_{i_1}^{k+1}, \ldots, x_{i_m}^{k+1}$ forms the basis, and the corresponding columns of T^{k+1} are all zero except for a 1 in the appropriate position. Occasionally, due to linear dependencies, it may not be possible to place the largest m components in the basis. In such a case, determine a basis out of the largest $m + 1$ components of x^{k+1} or, if necessary, $m + 2$, $m + 3$, etc. The magnitude of the components of x^k in the basis should be as great as possible. In summary, place the largest number of the largest components of x^{k+1} in the basis.†

As an example of the tableau-selection rule suppose the original constraints were

$$x_1 + 3x_2 \quad\quad -x_3 + 2x_4 = 7$$
$$x_1 - x_2 \quad\quad\quad\quad\quad = 1$$
$$x_1 \geq 0, x_2 \geq 0, x_3 \geq 0, x_4 \geq 0$$

If $x^k = (2, 1, 1, \frac{3}{2})^t$, then x_1 and x_4 would form the basis. The corresponding tableau is

$$T^k = \begin{bmatrix} 1 & -1 & 0 & 0 \\ 0 & 2 & -\frac{1}{2} & 1 \end{bmatrix}$$

where $x_B^k = (x_1^k, x_4^k)^t$, $B_1 = 1$, and $B_2 = 4$.

Should $x^k = (3, 2, 2, 0)^t$, then we may select $x_B^k = (x_1^k, x_2^k)^t$ with

$$T^k = \begin{bmatrix} 1 & 0 & -\frac{1}{4} & \frac{1}{2} \\ 0 & 1 & -\frac{1}{4} & \frac{1}{2} \end{bmatrix}$$

---

†For a different method of tableau selection see exercise 13.9.

In this case x_1 and x_3 could also have been selected to be basic with a corresponding tableau.

Note the right-hand side is immaterial.

Degeneracy

It is evident that should f be linear the convex-simplex method becomes the linear-simplex method. Furthermore, it is known that the linear-simplex method may not converge if a basic variable becomes zero. A similar difficulty could arise in the convex-simplex method. To avoid this degeneracy and possible nonconvergence, special procedures have been developed. However, these procedures do not seem necessary in practice. To simplify the presentation we instead pose the following nondegeneracy assumption which is analogous to the linear-simplex method nondegeneracy assumption.

Nondegeneracy assumption: The basic variables are always positive.

The nondegeneracy assumption is quite reasonable as the m largest components of x form the basis.

Proof of Convergence

We now can prove convergence of the procedure under the nondegeneracy assumption and the assumption that f be continuously differentiable. Convergence Theorem A will be used.

As is customary we assume all x^k to be in a compact set X so that condition 1 holds.

Let $x = z$ and define

$$Z(z) = f(x)$$

A point x^* is termed a solution if x^* satisfies the K-T conditions for problem (8.1). The following lemma will assist our proof. In the lemma a subscripted bracket connotes a component; thus $[b]_i = b_i$.

Lemma 8.1: *Consider problem (8.1) with f continuously differentiable. Let T^* be any linear programming tableau with b^* the corresponding right-hand side so that*

$$T^*x = b^*$$
$$x \geq 0$$

if and only if x is feasible. Also let x^* be a particular feasible point and

$$c^* = c(x^*) = [\nabla f(x^*) - (T^*)^t \nabla f(x^*)_B] \tag{8.22}$$

be its corresponding relative-cost vector. Suppose

$$c_{s_1}^* = \max \ \{c_i^* \,|\, i \in v\}$$

and

$$c_{s_2}^* x_{s_2}^* = \min \ \{c_i^* x_i^* \,|\, i \in v\}$$

Then if $c_{s_1}^* = c_{s_2}^* x_{s_2}^* = 0$, x^* is a solution.

PROOF. Because $c_{s_1}^* = c_{s_2}^* x_{s_2}^* = 0$, x^* satisfies

$$[\nabla f(x^*) - (T^*)^t \nabla f(x^*)_B]_i = \begin{cases} = 0 & \text{if} \quad x_i^* > 0 \\ \leq 0 & \text{if} \quad x_i^* = 0 \end{cases} \qquad (8.23)$$

In addition, since T^* is a tableau, a nonsingular $m \times m$ basis inverse matrix C^{-1} exists such that

$$T^* = C^{-1}A \qquad (8.24)$$

where A is the original constraint matrix.

Letting

$$u = (C^{-1})^t \nabla f(x^*)_B \qquad (8.25)$$

(8.23) becomes

$$[\nabla f(x^*) - A^t u]_i \begin{cases} = 0 & \text{if} \quad x_i^* > 0 \\ \leq 0 & \text{if} \quad x_i^* = 0 \end{cases} \qquad (8.26)$$

But (8.26) indicates that x^* satisfies the K-T conditions for 8.1.   ◆

Condition 2

Should $c_{s_1}^* = c_{s_2}^* x_{s_2}^* = 0$, then the algorithm terminates, and by Lemma 8.1, x^* is a solution, which proves condition 2(b) of Convergence Theorem A.

To establish 2(a), suppose $c_{s_1}^k > 0$ and $x_{s_1}^k$ is to be increased. Let case (a) occur and set

$$d^k = y^k - x^k$$

where d^k is defined in (8.19). Observe $d^k \neq 0$ because by the nondegeneracy assumption

$$\Delta^k > 0 \qquad (8.27)$$

Thus

$$d_i^k = 0 \qquad\qquad i \in N - s$$
$$d_s^k = \Delta^k$$
$$d_{B_i}^k = -t_{is}^k \Delta^k \qquad i \in \mu$$

Form

$$\nabla f(x^k)^t d^k = \sum_{i=1}^{n} \frac{\partial f(x^k)}{\partial x_i} d_i^k$$

$$= \Delta^k \left(\frac{\partial f(x^k)}{\partial x_s} - \sum_{i \in \mu} t_{is} \frac{\partial f(x^k)}{\partial x_{B_i}} \right) \qquad (8.28)$$

$$= \Delta^k c_s^k$$

using the definition of c_s^k.

As $\Delta^k > 0$ by (8.27), and $c_s^k > 0$ by assumption,

$$\nabla f(x^k)^t d^k > 0 \qquad\qquad (8.29)$$

But Equation (8.21) maximizes f in the direction d^k. Thus

$$f(x^{k+1}) > f(x^k) \qquad\qquad (8.30)$$

Similar proofs hold in the remaining cases and condition 2(a) holds.

Condition 3

Now we must verify condition 3.

Analysis of the algorithm reveals that we may break the algorithmic map A into four components

$$A = M^1 A^3 A^2 A^1$$

Here A^1 given x yields x again and the tableau T

$$A^1: x \longrightarrow (x, T)$$

The map A^2 from (x, T) gives the case a, b, or c, and the index of the nonbasic component to be changed, x_s, as well as (x, T) again

$$A^2: (x, T) \longrightarrow (x, T, s, \text{case})$$

Then A^3 calculates the direction d.

$$A^3: (x, T, s, \text{case}) \longrightarrow (x, d, \text{case})$$

Note that the tableau is no longer needed.

Finally, M^1 is the usual M^1 map, and it determines the successor point, but for this situation in addition to x and d, M^1 also depends upon the case. The case specifies the interval J, because J is [0, 1] in cases (a) and (c) but equals [0, a] in case (b).

$$M^1: (x, d, \text{case}) \longrightarrow (w)$$

where w is a successor to x.

Closedness of the Composition

We must prove that each of these maps is closed. The closedness of A will then follow from Corollary 4.2.1 because all maps are from compact sets to compact sets as we now show. We have assumed that all x are in a compact set X. There are only a finite number of tableaux so that all tableaux are also on a compact set. Consequently, because the map A^1 yields points (x, T), it is from a compact set to a compact set.

For A^2 the indices s clearly belong to a compact set, since there are only a finite number of them, while similarly the cases are in a compact set. Considering A^3, it maps to a compact set because the direction $d = y - x$, and all y and x are on the compact set X. Similarly M^1 is from a compact set to a compact set. Thus, once each component of A can be proved closed, A will be closed because all maps are from a compact set to a compact set.

Recall in proving closedness we may assume that x is not a solution and that all ties are arbitrarily broken.

Closedness of A¹

Let

(a) $x^k \longrightarrow x^\infty \qquad k \in \mathcal{K}$

and

(b) $T^k \longrightarrow T^\infty \qquad k \in \mathcal{K}$

where

(c) $(x^k, T^k) \in A^1(x^k) \qquad k \in \mathcal{K}$

It must be proved that

$$(x^\infty, T^\infty) \in A^1(x^\infty)$$

or equivalently that the tableau T^∞ could be generated at x^∞.

Because there are only a finite number of tableaux, a tableau T and a $\mathscr{K}^1 \subset \mathscr{K}$ must exist such that for $k \in \mathscr{K}^1$

$$T^k = T = T^\infty$$

At x^∞ the tableau T will also be a tableau with the largest number of the largest components of x^∞ in the basis. This follows because for $k \in \mathscr{K}^1$ if x_i^k is in the basis with respect to T and x_j^k is not

$$x_i^k \geq x_j^k$$

Then taking limits

$$x_i^\infty \geq x_j^\infty$$

Thus at x^∞, x_i^∞ can again be in the basis. Consequently, tableau T may be chosen at x^∞ for the tableau. Recall ties may be arbitrarily broken. Therefore, since $T = T^\infty$, A^1 is closed.

Closedness of A^2

Let
(a) $x^k \longrightarrow x^\infty$ and $T^k \longrightarrow T^\infty$ $k \in \mathscr{K}$
(b) $s^k \longrightarrow s^\infty$ and $\epsilon^k \longrightarrow \epsilon^\infty$ $k \in \mathscr{K}$
and
(c) $(x^k, T^k, s^k, \epsilon^k) \in A^2(x^k, T^k)$ $k \in \mathscr{K}$
where s^k is the component of the vector chosen to be altered (either increased or decreased) on the kth iteration, and ϵ^k indicates (a), (b), or (c) depending upon which case occurs.

We must establish that s^∞ and ϵ^∞ could be generated from $A^2(x^\infty, T^\infty)$, that is, at x^∞ with tableau T^∞, case ϵ^∞ occurs and component s^∞ can be altered.

Because x has only a finite number of components, a given one, say the sth, must have been chosen to be altered infinitely often. Also one of the three cases must have occurred infinitely often. Suppose that case is case (c); the other cases being simpler are left as an exercise. Finally, a given tableau T occurs infinitely often. Thus for some $\mathscr{K}^1 \subset \mathscr{K}$

$$s^k = s = s^\infty \qquad k \in \mathscr{K}^1$$
$$T^k = T = T^\infty \qquad k \in \mathscr{K}^1$$
$$\epsilon^k = c = \epsilon^\infty \qquad k \in \mathscr{K}^1$$

Writing case (c) out explicitly for $k \in \mathscr{K}^1$

$$c(x^k)_{s_1} \leq |c(x^k)_s x_s^k| \tag{8.31}$$

$$c(x^k)_{s_1} \geq c(x^k)_i \qquad i \in v \tag{8.31a}$$

and

$$|c(x^k)_s x_s^k| \geq |c(x^k)_i x_i^k| \qquad i \in v \tag{8.31b}$$

But because ∇f is continuous, $c(x)$ is continuous, and we may take limits of (8.31) yielding

$$c(x^\infty)_{s_1} \leq |c(x^\infty)_s x_s^\infty| \tag{8.32}$$

$$c(x^\infty)_{s_1} \geq c(x^\infty)_i \qquad i \in v \tag{8.32a}$$

and

$$|c(x^\infty)_s x_s^\infty| \geq |c(x^\infty)_i x_i^\infty| \qquad i \in v \tag{8.32b}$$

Hence, at x^∞ the component x_s is again a candidate to be decreased (as ties may be arbitrarily broken). Consequently, we may select $s^\infty = s$, and case (c) occurs. The map A^2 is closed.

Closedness of A^3

For A^3, as mentioned in the proof of A^2 above, we may assume

(a) $x^k \longrightarrow x^\infty \qquad k \in \mathcal{K}$

$\quad\ \ T^k = T = T^\infty \qquad k \in \mathcal{K}$

$\quad\ \ s^k = s = s^\infty \qquad k \in \mathcal{K}$

$\quad\ \ \epsilon^k = c = \epsilon^\infty \qquad k \in \mathcal{K}$

(b) $d^k \longrightarrow d^\infty \qquad k \in \mathcal{K}$

and

(c) $(x^k, d^k, \epsilon^k) \in A^3(x^k, T^k, s^k, \epsilon^k) \qquad k \in \mathcal{K}$

Closedness of A^3 will be verified by proving that it is possible to generate the direction d^∞ given the point x^∞, tableau T^∞, and the fact that x_s is to be decreased.

In case (c) either $\Delta^k = \Delta_1^k$ or Δ_2^k, $k \in \mathcal{K}$. Therefore there is a $\mathcal{K}^1 \subset \mathcal{K}$ such that one of the two occurs. Suppose

$$\Delta^k = \Delta_1^k \qquad k \in \mathcal{K}^1$$

(The argument for $\Delta^k = \Delta_2^k$ is similar.) Then by choice of Δ^k

$$\Delta^k = \frac{x_{B_r}^k}{t_{rs}} = \max \left\{ \frac{x_{B_i}^k}{t_{is}} \,\middle|\, t_{is} < 0 \right\} \tag{8.33}$$

Due to the fact that there are only a finite number of variables in the basis there is a $\mathcal{K}^2 \subset \mathcal{K}^1$ such that for $k \in \mathcal{K}^2$ a given $x_{B_i}^k/t_{is}$, say $x_{B_1}^k/t_{1,s}$, must achieve the maximum. Hence for $k \in \mathcal{K}^2$

$$\Delta^k = \frac{x_{B_1}^k}{t_{1,s}} = \max \ \left\{ \frac{x_{B_i}^k}{t_{is}} \middle| t_{is} < 0 \right\} \tag{8.34}$$

Taking limits of (8.34), we arrive at

$$\Delta^k \longrightarrow \Delta^\infty = \frac{x_{B_1}^\infty}{t_{1,s}} = \max \ \left\{ \frac{x_{B_i}^\infty}{t_{is}} \middle| t_{is} < 0 \right\} \tag{8.35}$$

But then via Equation (8.19) for $k \in \mathcal{K}^2$

$$y^k \longrightarrow y^\infty$$

where

$$\begin{aligned} y_i^\infty &= x_i^\infty & i \in N - s \\ y_s^\infty &= x_s^\infty + \Delta^\infty \end{aligned} \tag{8.36}$$

and

$$y_{B_i}^\infty = x_{B_i}^\infty - t_{is}\Delta^\infty \qquad i \in \mu$$

Thus by the construction of d^k

$$d^k \longrightarrow d^\infty = y^\infty - x^\infty$$

But at x^∞, since case (c) occurs and from (8.35) and (8.36), we see that d^∞ can be chosen as the direction. Consequently, A^3 is closed.

Finally, the map M^1 is closed for any of the three cases by Lemma 5.1.

In summary, each of the component maps of A is closed. Condition 3 is verified and the convex-simplex method converges.

An Example

Consider the problem

$$\begin{aligned} \min \quad & (x_1)^2 + x_2 x_3 + (x_4)^2 \\ \text{subject to} \quad & \tfrac{1}{4}x_1 + x_2 + x_3 + \tfrac{1}{2}x_4 & \geq 4 \\ & \tfrac{1}{4}x_1 + x_2 & \geq 2 \\ & x_1 \geq 0 \quad x_2 \geq 0 \quad x_3 \geq 0 \quad x_4 \geq 0 \end{aligned}$$

Observe the minimization instead of maximization.

After addition of slack variables the constraints become

$$\tfrac{1}{4}x_1 + x_2 + x_3 + \tfrac{1}{2}x_4 - x_5 \quad\quad = 4$$
$$\tfrac{1}{4}x_1 + x_2 \quad\quad\quad\quad\quad - x_6 = 2$$
$$x_1 \geq 0 \quad\quad x_2 \geq 0 \quad\quad x_3 \geq 0 \quad\quad x_4 \geq 0 \quad\quad x_5 \geq 0 \quad\quad x_6 \geq 0$$

Let us suppose that several iterations have already taken place, and that $x^3 = (0, 2, 3, 0, 1, 0)$ has just been generated. For simplicity transposes will be omitted. The x^3 and its corresponding tableau are

x^3	0	2	3	0	1	0
T^3	$\tfrac{1}{4}$	1	0	0	0	-1
	0	0	1	$\tfrac{1}{2}$	-1	1

Here

$$x_B^3 = \begin{pmatrix} x_{B_1} = x_2 = 2 \\ x_{B_2} = x_3 = 3 \end{pmatrix}$$

Step 1:

$$\nabla f(x^3) = (2x_1, x_3, x_2, 2x_4, 0, 0) = (0, 3, 2, 0, 0, 0)$$
$$\nabla f(x^3)_B = (x_3, x_2) = (3, 2)$$

$$c^3 = [\nabla f(x^3) - (T^3)^t \nabla f(x^3)_B] = \begin{pmatrix} 0 & -\tfrac{3}{4} \\ 3 & -3 \\ 2 & -2 \\ 0 & -1 \\ 0 & +2 \\ 0 & +1 \end{pmatrix} = \begin{pmatrix} -\tfrac{3}{4} \\ 0 \\ 0 \\ -1 \\ +2 \\ +1 \end{pmatrix}$$

Because the problem is to minimize we interchange the min and max operations for calculating x_s.

$$c_{s_1}^3 = \min \ \{c_i^3\} = -1 \quad\quad s_1 = 4$$
$$c_{s_2}^3 x_{s_2}^3 = \max \ \{c_i^3 x_i^3\} = 2 \quad\quad s_2 = 5$$

Step 2: $c_5^3 x_5^3 > |c_4^3|$. Therefore decrease x_5. Case (c) has occurred.

Step 3 case (c)

$$\Delta_1^3 = \left\{ \frac{3}{-1} \right\} = -3$$
$$\Delta_2^3 = -1$$

so that

$$\Delta^3 = \max \quad \{-1, -3\} = -1$$

$$y^3 = \begin{pmatrix} 0 \\ 2 \\ 2 \\ 0 \\ 0 \\ 0 \end{pmatrix}$$

$$d^3 = \begin{pmatrix} 0 \\ 0 \\ -1 \\ 0 \\ -1 \\ 0 \end{pmatrix}$$

To find x^4 calculate

$$\min_{0 \le \tau \le 1} \; f(x^3 + \tau d^3) = \min_{0 \le \tau \le 1} \; 2(3 - \tau)$$

Solving, we find $\tau^3 = 1$ and $x^4 = y^3$.

Iteration 4. No pivoting is necessary

x^4	0	2	2	0	0	0
T^4	$\frac{1}{4}$	1	0	0	0	-1
	0	0	1	$\frac{1}{2}$	-1	1

and $x_B^4 = \begin{pmatrix} x_{B_1} = x_2 = 2 \\ x_{B_2} = x_3 = 2 \end{pmatrix}$

$$\nabla f(x^4) = (0, 2, 2, 0, 0, 0) \qquad \nabla f(x^4)_B = (2, 2)$$

Then

$$c^4 = \begin{pmatrix} 0 & -\frac{1}{2} \\ 2 & -2 \\ 2 & -2 \\ 0 & -1 \\ 0 & +2 \\ 0 & 0 \end{pmatrix} = \begin{pmatrix} -\frac{1}{2} \\ 0 \\ 0 \\ -1 \\ +2 \\ 0 \end{pmatrix}$$

Step 2

$$c_{s_1}^4 = -1 \qquad s_1 = 4 \qquad c_{s_2}^4 x_{s_2}^4 = 0$$

We increase x_4.

Step 3 case (a)

$$\Delta = \left\{ \frac{2}{1/2} \right\} = 4$$

$$y^4 = \begin{pmatrix} 0 \\ 2 \\ 0 \\ 4 \\ 0 \\ 0 \end{pmatrix}$$

Solving $x^5 = (0, 2, \frac{7}{4}, \frac{1}{2}, 0, 0)$.

The convex-simplex method continues in this manner.

8.2.1. Some Geometrical Considerations in the Convex-Simplex Method

The convex-simplex method attempts to move, as much as possible, like the linear-simplex method. The linear-simplex method at a vertex of the

The linear-simplex method at vertex a moves either along the edge ab or edge ad.

The convex-simplex method at point x^k figuratively moves vertex a into the center of the feasible region. Vertex a is the reference vertex.

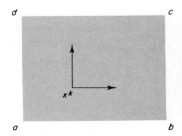

Fig. 8.2.

Reference vertices in the convex-simplex method. The feasible region is the rectangle a, b, c, and d.

feasible region selects an edge and moves along this edge until the next vertex is encountered. For the convex-simplex method the point x^k may not be at a vertex but might be inside the feasible region. Nevertheless a basis is chosen. Intuitively, this basis determines a reference vertex, which is the vertex that would be determined were all the nonbasic variables zero. It is with regard to this reference vertex that the convex-simplex method moves, and the movement is parallel to an edge forming the reference vertex (see Fig. 8.2). Of course, when a nonbasic variable is decreased, the movement might be opposite to the direction arrows indicated in the figure.

8.3. Concave Maximization by Manifold Suboptimization

The method presented in this section is for problem (8.1) with f concave. It proceeds by solving subproblems, each subproblem consisting of only some of the constraints of problem (8.1), and convergence occurs after the algorithm solves only a finite number of these subproblems. However, the algorithm is not a finite process because it may not be possible to solve the subproblems in a finite number of iterations. A special case in which finite convergence is possible will be treated in the next chapter.

Motivation

To introduce the method consider problem (8.1) where f is a concave continuously differentiable function, and let $x^* = (x_1^*, \ldots, x_n^*)$ be optimal. We will write the K-T conditions for this problem in a special manner. First define

$$B(x) = \{i \,|\, x_i = 0\}$$

Then $B(x)$ is the set of indices for which the corresponding components of the point x is zero.

Using $B(x^*)$ the K-T conditions become:

There exist x^* feasible, multipliers u_i, $i = 1, \ldots, m$, unconstrained, and $\lambda_i^* \geq 0$, $i = 1, \ldots, n$, such that [K-T condition (2)]

$$\lambda_i^* = 0 \qquad i \notin B(x^*) \tag{8.37}$$

and [K-T condition (3)]

$$\nabla f(x^*) + A^t u^* + \lambda^* = 0 \tag{8.38}$$

Recall (exercise 2.13) that with linear constraints the constraint qualification automatically holds. Moreover, because f is concave, the K-T conditions are also sufficient.

Elimination of Inequalities

Now suppose that we consider an intimately related problem

$$\text{max} \quad f(x)$$
$$\text{subject to} \quad Ax = b \tag{8.39}$$
$$x_i = 0 \quad i \in B(x^*)$$

For simplicity assume the optimal solution to problem (8.39) is unique. Then the K-T conditions that x' be optimal for problem (8.39) may be written as follows:

There exist x' feasible, unconstrained multipliers u_i', $i = 1, \ldots, m$, and unconstrained λ_i', $i = 1, \ldots, n$, such that [K-T condition (2)]

$$\lambda_i' = 0 \quad i \notin B(x^*) \tag{8.40}$$

and [K-T condition (3)]

$$\nabla f(x') + A^t u' + \lambda' = 0 \tag{8.41}$$

By the same reasoning as above the K-T conditions are both necessary and sufficient.

Examination of the K-T conditions reveals that x^*, the optimal point for problem (8.1), is also optimal for problem (8.39). By assumption problem (8.39) has a unique optimal solution. Consequently, x^* is the only optimal solution to problem (8.39). We are thus led to the following conclusion. Solving problem (8.39) will also solve problem (8.1).

The key import of this result is as follows. If we can determine the constraints that are active at the final solution of problem (8.1), we can solve a simpler problem (8.39) to determine the optimal solution to problem (8.1). The optimization technique to be discussed is based upon this approach and presents an adaptive method for determining the constraints that are active at the optimal solution. The method is extremely powerful if problems such as (8.39) are easily solved (exercise 8.8). One such case when f is quadratic is discussed in the next chapter.

Manifolds

Problem (8.39) has the form of maximizing f over the the set of solutions to a set of linear equations. The set of solutions of a set of linear equations is a (linear) manifold. Thus problem (8.39) has the form of maximizing f over a manifold.

Let us examine manifolds in more detail. It was mentioned in Chap. 6 that a manifold is similar to a linear subspace of E^n except that, unlike a subspace, the manifold need not go through the origin. Examples of manifolds in E^n include E^n itself, a point, and a hyperplane. Important to our discussion is the dimension of a manifold. In terms of solutions to systems of linear equations, the dimension is the largest number of linearly independent vectors we may place in the solution set. For example, the dimension of E^n is n, a hyperplane has dimension $n - 1$, while a point has dimension zero. Moreover, suppose we are given an $r \times n$ matrix B whose rows are linearly independent. Then the manifold that is the set of solutions to the equations

$$Bx = c \tag{8.42}$$

has dimension $n - r$. In the vacuous case when $r = 0$ the dimension is n. Clearly, the addition of a linearly independent equation to (8.42) reduces the dimension of the manifold.

The Manifold-Suboptimization Procedure

The manifold-suboptimization procedure operates essentially as follows. Suppose a feasible point x^1 for problem (8.1) is given. Then we solve the problem

$$\max \quad f(x)$$
$$\text{subject to} \quad Ax = b$$
$$x_i = 0 \qquad i \in B(x^1)$$

for an optimal point y^1.

If in the first case y^1 is feasible for problem (8.1), then we go to y^1 and let $x^2 = y^1$. If y^1 is also optimal for problem (8.1), the procedure stops, while should y^1 not be optimal, a new problem is determined by a method to be described shortly, and the procedure continues.

Suppose, on the other hand, that y^1 is not feasible. We then proceed from x^1 directly toward y^1 until a boundary point is reached. That boundary point is x^2. In addition, because f is concave, x^2 also satisfies

$$x^2 \in M^1(x^1, d^1)$$

where

$$d^1 = x^2 - x^1$$

and $J = [0, 1]$. This follows because f concave implies f is an increasing function as we move from x^1 to y^1.

The procedure then continues from x^2.

Assumptions

Before stating the procedure precisely we must develop a lemma and some definitions. The lemma indicates how to obtain a better point if, after optimizing f in a manifold, the resulting point is not optimal for problem (8.1). For notational simplicity define

$$Y_B = \{x \mid Ax = b,\ x_i = 0,\ i \in B\} \tag{8.43}$$

where B is some subset of the integers $\{1, \ldots, n\}$. Observe, problem (8.39) may be reformulated as

$$\max\ \{f(x) \mid x \in Y_{B(x^*)}\}$$

Uniqueness

A uniqueness property is also required. Suppose y is optimal for the problem

$$\max\ \{f(x) \mid x \in Y_B\} \tag{8.44}$$

Part of the K-T conditions for this problem, where $\lambda_i = 0,\ i \notin B$, are

$$\nabla f(y) + A^t u + \lambda = 0 \tag{8.45}$$

We may reformulate (8.45) as

$$\nabla f(y) + \sum_{i=1}^{m} u_i a^i + \sum_{i \in B} \lambda_i c^i = 0 \tag{8.46}$$

Here $\nabla f(y)$ is being expressed as a linear combination of both the rows $a^i,\ i = 1, \ldots, m$, of A, and the unit coordinate vectors $c^i,\ i \in B$. Recall a unit coordinate vector contains zeroes except for a one in the ith position.

The uniqueness property may now be defined. The point y satisfies the **uniqueness property** if $\nabla f(y)$ is expressed as a unique linear combination of the rows of A and the unit coordinate vectors $c^i,\ i \in B$. In practice the uniqueness property nearly always holds.

With these definitions we pose the following lemma.

Lemma 8.2: *Let y be optimal for*

$$\max \quad \{f(x) \mid x \in Y_B\} \tag{8.47}$$

and suppose y satisfies the uniqueness property. Letting part of the K-T conditions for (8.47) be

$$\nabla f(y) + \sum_{i=1}^{m} u_i a^i + \sum_{i \in B} \lambda_i c^i = 0 \tag{8.48}$$

assume for some $j \in B$ that

$$\lambda_j < 0$$

Then, if y' is optimal for

$$\max \quad \{f(x) \mid x \in Y_{B'}\} \tag{8.49}$$

where $B' = B - \{j\}$ (i.e., the set B except for $\{j\}$),

$$f(y') > f(y)$$

PROOF: Note that since $\lambda_j < 0$, y satisfies the K-T conditions for the problem

$$\begin{aligned}
\max \quad & f(x) \\
\text{subject to} \quad & Ax = b \\
& x_i = 0 \qquad i \in B' \\
& x_j \leq 0
\end{aligned} \tag{8.50}$$

and since the K-T conditions are sufficient with f concave and linear constraints, y is optimal for problem (8.50).

By the uniqueness property there cannot exist a set of u and λ_i, $i \in B$, that satisfy (8.48) with $\lambda_j \geq 0$. Therefore, the K-T conditions do not hold at y for the problem

$$\begin{aligned}
\max \quad & f(x) \\
\text{subject to} \quad & Ax = b \\
& x_i = 0 \qquad i \in B' \\
& x_j \geq 0
\end{aligned} \tag{8.51}$$

and y is not optimal for problem (8.51).

Letting y^* be optimal for problem (8.51)

$$f(y^*) > f(y)$$

But because problem (8.49) is less constrained than problem (8.51)

$$f(y') \geq f(y^*) > f(y) \quad \blacklozenge \qquad (8.52)$$

Comment

Briefly, the lemma states that if y is optimal for problem (8.47) with $\lambda_j < 0$, then by releasing the "tight" constraint $x_j = 0$, a higher value of f can be obtained.

We now pose a precise statement of the algorithm for maximizing a concave function via manifold suboptimization.

Statement of the Algorithm

Suppose a point x^1 feasible for problem (8.1) is given. Let $B^1 = B(x^1)$. Go to step I with $k = 1$.

Step I. Determine y^k to

$$\max \ \{f(x) \,|\, x \in Y_{B^k}\} \qquad (8.53)$$

Case 1: y^k is feasible for problem (8.1).
Using the K-T multipliers for problem (8.53) determine

$$\lambda_j = \min \ \{\lambda_i\}$$

If $\lambda_j \geq 0$, terminate; y^k is optimal for problem (8.1).
If $\lambda_j < 0$, define

$$B^{k+1} = B^k - \{j\}$$

and

$$x^{k+1} = y^k$$

Go to step I with $k + 1$ replacing k.

Case 2: y^k is infeasible.
Let x^{k+1} be the feasible point closest to y^k on the line between x^k and y^k.

Set

$$B^{k+1} = B(x^{k+1})$$

Go to step I with $k + 1$ replacing k.

Proof of Convergence in a Finite Number of Steps

Clearly, if the algorithm terminates, y^k must be optimal as the K-T conditions are satisfied.

We now demonstrate that termination must occur after a finite number of iterations. The assumptions required are that f be concave and continuously differentiable and that each problem (8.53) has a solution which satisfies the uniqueness property.

First, we show whenever case 1 occurs and there is no termination that the objective function must strictly increase on the next step. Specifically, suppose y^{k-1} is feasible but not optimal so that $x^k = y^{k-1}$. It will be shown that

$$f(x^{k+1}) > f(x^k)$$

If y^k is also feasible, then Lemma 8.2 ensures the result as $x^{k+1} = y^k$. If y^k is not feasible, Lemma 8.2 does provide that

$$f(y^k) > f(x^k)$$

But by concavity of f any point between x^k and y^k is larger than x^k. Hence

$$f(x^{k+1}) > f(x^k)$$

(Here we suppose there is no degeneracy and $x^{k+1} \neq x^k$.)

Thus it has been proved that when case 1 occurs

$$f(x^{k+1}) > f(x^k) \tag{8.54}$$

We now can establish that case 1 can only occur a finite number of times. If case 1 occurs, x^{k+1} is set equal to y^k, the solution of a given subproblem. Moreover, the objective function is monotonic and from (8.54) that subproblem cannot recur. Yet due to the form of the subproblems, there are only a finite number of them. Consequently, case 1 can only recur a finite number of times without termination.

To prove the result we need only establish that case 2 can occur at most a finite number of times in succession without an occurrence of case 1. Then because case 1 occurs only a finite number of times, termination occurs in a finite number of steps.

Suppose case 2 occurs so that y^k is infeasible. Then the entire line segment between x^k and y^k is in the manifold Y_{B^k}. Moreover, x^{k+1} is in Y_{B^k}. But also at x^{k+1} an additional variable $x_r^{k+1} = 0$ for some $r \notin B^k$. Thus

$$B^{k+1} \supset B^k \cup \{r\}$$

Consequently, the dimension of the manifold $Y_{B^{k+1}}$ is less than Y_{B^k} because at least one additional constraint, namely, $x_r = 0$, has been added.

It has just been verified that every time case 2 occurs the dimension of the manifold used in the next subproblem decreases. Should case 2 occur sufficiently often in succession a zero dimensional manifold must result. But a zero dimensional manifold is a point, say x^k. The subproblem of maximizing f over the manifold consisting of x^k alone must yield $y^k = x^k$. However, as x^k is feasible, case 1 occurs. Therefore, since there is only a finite number of dimensions, and because each time case 2 occurs the dimension of the manifold decreases, case 1 must occur after at most a finite number of repetitions of case 2.

In conclusion, because case 1 without termination occurs after a finite number of case 2 repetitions and because case 1 can only eventuate a finite number of times, termination must result after a finite number of repetitions of step I.

Geometrical Behavior

Geometrically, in the manifold procedure we move toward a point that optimizes the objective function in a manifold. If the manifold optimal point is feasible but not optimal for problem (8.1), then the new point is the manifold optimal point, and a manifold of one large dimension is considered by dropping the requirement that $x_j = 0$ where $\lambda_j < 0$. Should the manifold optimal point be infeasible, then the feasible point closest to the manifold optimal point on the line between the present point and the manifold optimal point is taken as the new point. At the new point an additional component of x becomes zero, say $x_r = 0$. We then form a new manifold by adding the constraint $x_r = 0$ to the previous manifold, and move toward the manifold optimal point for the new manifold. The procedure continues in this manner.

EXERCISES

8.1. For the linear-approximation method prove that the K-T conditions are satisfied if x^* is a solution.

8.2. Show that the linear-approximation method converges to an optimal point if f is pseudoconcave.

8.3. Prove for f concave in the linear-approximation method that $f(y^*)$ is an upper bound for the solution value $f(x^*)$. How might this information be helpful in computation?

8.4. For the linear-approximation method define the function

$$j(x) = \max \quad \{\nabla f(x)^t(y - x) \,|\, y \in F\}$$

and the map

$$H(x) = \{y \,|\, j(x) = \nabla f(x)^t(y - x), y \in F\}$$

Then the map D in $A = M^1 D$ can be expressed as

$$D(x) = \{(x, d) \,|\, d = y - x, y \in H(x)\}$$

Prove D is closed using Lemma 7.3.

8.5. *A Second-Order Method.* Suppose for problem (8.1) that f has continuous second partial derivatives. Consider a method that iterates the same as the linear-approximation method except that the subproblem for y given x is

$$\max \quad \nabla f(x)^t y + \tfrac{1}{2} y^t H(x) y$$
$$Ay = b$$
$$y \geq 0$$

where $H(x)$ is the Hessian matrix.

Prove convergence of this method. What assumptions are needed? What is a solution point?

8.6. Give a proof of the convex-simplex method for all cases that could arise.

8.7. In the convex-simplex method consider an alternative method for selecting the variable to change. Let $c_{s_1}^k$ be defined as before but choose

$$c_{s_2}^k = \min \quad \{c_i^k \,|\, x_i^k > 0, i \in v\}$$

The rules then are

If $\quad c_{s_1}^k \geq |c_{s_2}^k| \quad$ increase $\quad x_{s_1}^k$
If $\quad c_{s_1}^k \leq |c_{s_2}^k| \quad$ decrease $\quad x_{s_2}^k$

Verify that the convergence proof given in the text does not hold for these rules.

8.8. *Maximizing a Function in a Linear Manifold.* Let f be a continuously differentiable concave function, c an n vector, and B an $r \times n$ matrix $r \leq n$

with linearly independent rows. Suppose we must find the optimal solution to the problem

$$\max \quad f(x)$$
$$\text{subject to} \quad Bx = c$$

(a) Show that the K-T conditions for x^* to be optimal for this problem are

$$Bx^* = c$$
$$\nabla f(x^*) + B^t u = 0$$

where u is an unconstrained r vector.

Define the $n \times n$ matrix

$$I - B^t(BB^t)^{-1}B$$

where $(BB^t)^{-1}$ is the inverse of an $r \times r$ matrix (BB^t).

(b) Let x be a point feasible for the problem, and suppose

$$[I - B^t(BB^t)^{-1}B]\nabla f(x) = 0$$

Prove that x is optimal for the problem.

(c) Prove

$$[I - B^t(BB^t)^{-1}B]^t[I - B^t(BB^t)^{-1}B] = [I - B^t(BB^t)^{-1}B]$$

Hint:

$$[B^t(BB^t)^{-1}B]^t = [B^t(BB^t)^{-1}B]$$

Now define

$$d = [I - B^t(BB^t)^{-1}B]\nabla f(x)$$

(d) Prove

$$\nabla f(x)^t d = d^t d$$

(e) If x is not optimal for the problem, then show that

$$\nabla f(x)^t d > 0$$

(f) Let x be feasible for the problem. Prove $x + \tau d$ is also feasible for all τ.

(g) Prove that the following algorithm for the problem converges

$$A = M^1 D$$

where for d defined above

$$D(x) = (x, d)$$

and $J = [0, \alpha]$ for α positive.

Also the initial point x^1 is given as feasible.

8.9. Show that the manifold-optimization procedure also converges for f pseudoconcave.

8.10. In the convex-simplex method prove that if f is linear the relative costs obtained by Equation (8.13) are the same as those calculated by the linear-simplex method.

NOTES AND REFERENCES

§8.1. This algorithm was suggested by FRANK and WOLFE, although the proof via the convergence theory is new.

§8.2. The convex-simplex method was introduced in ZANGWILL (1967g) and has been extended and revealed to be decomposable as well as applicable to many special matrix forms by RUTENBERG. The convex-simplex method is related to an algorithm developed by Wolfe, see WOLFE (1962); LHERMITTE and BESSIERE; and FAURE and HUARD [see also ROSEN (1960) and BEALE (1955)].

§8.3. The discussion is based upon work by ZANGWILL (1967h).

Exercise 8.8 gives a brief introduction to the topic of generalized inverses, see PENROSE. Generalized inverses extend the notion of matrix invertability to matrices that do not possess an ordinary inverse.

ADDITIONAL COMMENTS

For special structures, the problem of maximizing a function subject to linear constraints simplifies. Such simplifications occur for a separable objective function, see CHARNES and LEMKE; or MILLER. Simplifications also arise due to network formulations, see BEALE (1959); HU; and ZANGWILL (1966b).

9 Acceleration Techniques

and the

Quadratic Programming Problem

The previous chapter presented the convex-simplex method and the manifold-suboptimization method for problem (8.1), the nonlinear programming (NLP) problem with linear constraints. In this chapter we specialize each of these methods to solve the quadratic programming (QP) problem:

$$\max \quad q^t x + \tfrac{1}{2} x^t Q x$$
$$Ax = b \qquad\qquad (9.1)$$
$$x \geq 0$$

Conjugate directions will be incorporated into the convex-simplex method to form a new method called the convex-simplex-method-conjugate-direction (CSM-CD) algorithm. This algorithm not only solves problem (9.1) in a finite number of steps, but also it is quite helpful for a general objective function f, because in such cases the conjugate directions tend to accelerate convergence.

The manifold-suboptimization method for problem (9.1) specializes to a pivot algorithm that operates by pivoting in appropriate tableaux. Unlike the CSM-CD method, however, the pivot algorithm is not useful for general

f. Nevertheless, it is quite efficient for the QP problem when Q is negative semidefinite.

9.1. QUADRATIC CONVERGENCE OF THE CONVEX-SIMPLEX METHOD

We will now show how to couple the convex-simplex method and a conjugate-direction method by developing the CSM-CD method. The conjugate directions will enable the convex-simplex method to converge in a finite number of steps to a point at which the K-T conditions hold for a general quadratic objective function. We will require no special form of the Q matrix. Moreover, with a nonquadratic objective function, the conjugate directions will assist us by exploiting quadratic approximations and, hopefully, by speeding convergence. The conjugate-direction method to be used will be the procedure for maximizing a quadratic function: 2, procedure 2 of Chap. 6. For the spacer step of procedure 2 we employ a convex-simplex method step. Then for general f Convergence Theorem B will ensure convergence because the convex-simplex method will be the basic algorithm. A knowledge of both the convex-simplex method and procedure 2 is assumed.

The Map M^2

To present the CSM-CD method, we must introduce the map M^2. Recall F denotes the feasible region. The map $M^2: E^{2n} \longrightarrow E^n$, given $x \in F$ and a direction d, determines

$$y \in M^2(x, d) \tag{9.2}$$

by

$$f(y) = \max \{f(x + \tau d) \mid \tau \in J, x + \tau d \in F\}$$

where

$$y = x + \tau^0 d$$

for J an interval, $\tau^0 \in J$, and $y \in F$.

Essentially, $M^2(x, d)$ maximizes f in the direction d from x but only over the portion of $x + \tau d$ for $\tau \in J$ that is in the feasible region. Note that $y \in M^2$ must also be feasible.

Relation Between M^1 and M^2

Suppose $J = (-\infty, +\infty)$, then $M^2(x, d)$ maximizes f on the portion of the line through x formed by the direction d that is in the feasible region. For example, consider problem (8.1) with x feasible. Let τ_1 be the smallest τ and τ_2 be the largest τ such that $x + \tau d$ is feasible. Generally, $\tau_1 \leq 0$. Then because the constraints are linear, for all τ, $\tau_1 \leq \tau \leq \tau_2$,

$$x + \tau d \in F$$

At τ_2 a variable, say x_i, will become zero, and increasing τ above τ_2 would drive x_i negative. Similarly, decreasing τ below τ_1 would drive a variable negative. Therefore, calculating

$$y \in M^2(x, d) \quad \text{where} \quad J = (-\infty, +\infty)$$

is equivalent to determining

$$y \in M^1(x, d) \quad \text{with} \quad J = [\tau_1, \tau_2]$$

Of course, should $\tau_1 = -\infty$ and $\tau_2 = +\infty$,

$$M^1(x, d) = M^2(x, d) \tag{9.3}$$

Now suppose for $y \in M^1(x, d)$ that

$$y = x + \tau^0 d \quad \text{with} \quad \tau_1 \leq \tau^0 \leq \tau_2$$

Then the maximum of f on the entire line has been achieved within the feasible region. More specifically in this case

$$y \in M^2(x, d) \quad \text{with} \quad J = (-\infty, +\infty) \tag{9.4}$$

has the property that

$$y \in M^1(x, d) \quad \text{with} \quad J = (-\infty, +\infty) \tag{9.5}$$

For the procedure to be described it will be important to know when

$$y \in M^2(x, d) \quad \text{with} \quad J = (-\infty, +\infty)$$

implies

$$y \in M^1(x, d) \quad \text{with} \quad J = (-\infty, +\infty)$$

Then y both maximizes f on the entire line and the portion of the line that is feasible.

Development of CSM-CD Procedure

The CSM-CD procedure is based upon a modification of the procedure for maximizing a quadratic function: 2. For the spacer step, step (a) in procedure 2, we utilize a convex-simplex method step. The notations x_i^j and d_i^j are, respectively, points and directions in E^n.

Roughly speaking, the CSM-CD procedure follows procedure 2 as long as procedure 2 can generate feasible points. Whenever procedure 2 cannot generate a feasible point, we stop using it and employ instead the map M^2 to ensure feasibility. Then from the feasible point generated by M^2 the CSM-CD procedure, after an adjustment step, restarts procedure 2. Procedure 2 is employed as much as possible in an effort to build up conjugate directions. In effect, whenever procedure 2 generates an infeasible point, we substitute a feasible point and recommence procedure 2.

Instead of employing map M^1 as in procedure 2, the CSM-CD procedure uses M^2 with $J = (-\infty, +\infty)$. This modification ensures all points are feasible. If the points generated by M^2 are also in M^1 then steps 0 and 1 of the CSM-CD procedure are precisely procedure 2. Hence, if f is quadratic, conjugate directions will be generated, and the procedure will terminate in a finite number of iterations. Whenever a point in M^2 is not in M^1, then that implies that procedure 2 could not generate a feasible point and we, after an adjustment step, restart from step 0 but with the current point. Every time step 0 restarts we are actually recommencing procedure 2, and our effort to construct conjugate directions is begun again.

Letting the matrix A be $m \times n$, define $r = n - m$. Then because there are r free variables in the problem, we need only generate r conjugate directions.

To initiate the CSM-CD procedure a feasible point x^0 will be given. Step 0 in the CSM-CD procedure employs directions $d_i^1 = 0$, $i = 1, \ldots,$ $r - 1$. For d_r^1 the direction the convex-simplex method would determine at x^0 will be utilized. Then the first point $x_{r+2}^0 \in M^2(x^0, d_r^1)$. Observe x_{r+2}^0 is the same as if the convex-simplex method were applied to x^0. If $x_{r+2}^0 \in M^1(x^0, d_r^1)$ for $J = (-\infty, +\infty)$, then, at least so far, our CSM-CD procedure would be behaving identically to procedure 2. If $x_{r+2}^0 \notin M^1(x^0, d_r^1)$, the point generated by procedure 2 would be infeasible. In this case we repeat step 0 after an adjustment step and recommence our effort to construct conjugate directions.

Step 2 of the method serves as the adjustment step. It calculates $[c^j]_l [x^j]_l$ for all $l = 1, \ldots, n$ in which $[c^j]_l$ is the lth component of the relative-cost vector c^j and $[x^j]_l$ is the lth component of x^j. If $[c^j]_l [x^j]_l < 0$, then we decrease $[x]_l$, adjusting only basic variables until either f no longer increases, or a variable becomes zero, whichever occurs first. Thus the adjustment of $[x]_l$ is precisely case (c) in the convex-simplex method with $[x]_l$ playing the role of $[x]_s$.

The purpose of step 2 is to decrease positive nonbasic variables. With this accomplished, it is hoped that the basic variables will be as large as possible, thereby allowing larger changes in the basic variables before driving a basic variable to zero.

As mentioned above, x_i^j will be a point in E^n while $[x_j]_l$ denotes the lth component of x^j, and the d_i^j are directions in E^n. Also let m and n be determined by the $m \times n$ constraint matrix A. For simplicity, we suppose the feasible region F to be compact, then the maxima for M^1 will exist (see also exercise 9.1).

A CSM-CD Algorithm

In the procedure for both M^1 and M^2, $J = (-\infty, +\infty)$. An initial basic feasible point x^0 and corresponding tableau is given. Define $r = n - m$. Set $j = 0$, and go to step 0.

Step 0: Let $j + 1$ replace j. x^{j-1} is given. Set $d_i^j = 0$, $i = 1, \ldots, r - 1$. Let x_{r+2}^{j-1} be the point and d_r^j be the direction determined by the convex-simplex method at x^{j-1}. Pivot, placing the largest number of largest variables in the basis. If

$$x_{r+2}^{j-1} \in M^1(x^{j-1}, d_r^j)$$

go to step 1. Otherwise let $x^j = x_{r+2}^{j-1}$, and go to step 2.

Step 1: The point x_{r+2}^{j-1} and directions d_i^j, $i = 1, \ldots, r$, are given.
(a) Spacer step using a convex-simplex-method step on x_{r+2}^{j-1} to obtain x_1^j. Pivot to place the largest number of largest components of x_1^j in the basis.
(b) For $i = 1, \ldots, r$ do the following:
  Calculate $x_{i+1}^j \in M^2(x_i^j, d_i^j)$. If $x_{i+1}^j \in M^1(x_i^j, d_i^j)$, continue, otherwise go to step 2 with $x^{j+1} = x_{i+1}^j$, and replace j with $j + 1$.
(c) Set $d_{r+1}^j = x_{r+1}^j - x_{r+2}^{j-1}$. Calculate $x_{r+2}^j \in M^2(x_{r+1}^j, d_{r+1}^j)$. If $x_{r+2}^j \in M^1(x_{r+1}^j, d_{r+1}^j)$, go to d, otherwise go to step 2 with $x^{j+1} = x_{r+2}^j$, and replace j by $j + 1$.
(d) Set $d_i^{j+1} = d_{i+1}^j$, $i = 1, \ldots, r$. Go to step 1 with $j + 1$ replacing j.

Step 2: x^j is given. Pivot, placing the largest number of largest variables in the basis. Set $l = 1$.
(a) If $l \leq n$ continue, otherwise go to step 0.
  If $[c^j]_l[x^j]_l < 0$, decrease $[x]_l$ adjusting only basic variables until f no longer increases, $[x]_l$ becomes zero, or a basic variable becomes zero, whichever occurs first. Call the corresponding x, x^{j+1}. Replace j by $j + 1$. Pivot as in the convex-simplex method letting the largest number of largest variables be in the basis. Calculate the corresponding c^j. Increase l to $l + 1$, and go to (a).
  If $[c^j]_l[x^j]_l \geq 0$, set l to $l + 1$, and go to (a).

Comments

The behavior of step 0 is analogous to the initialization step of procedure 2.

For step 1(b), after calculating $x_{i+1}^j \in M^2$, check to see if $x_{i+1}^j \in M^1$. If so, calculate $x_{i+2}^j \in M^2(x_{i+1}^j, d_{i+1}^j)$ and continue the process in the same manner for x_{i+2}^j. If $x_{i+1}^j \notin M^1$, go to step 2 with $x^{j+1} = x_{i+1}^j$ and increase j to $j + 1$ (see also exercise 9.2).

For step 0 the point x_{r+2}^{j-1} calculated by the convex-simplex method can also be written

$$x_{r+2}^{j-1} \in M^2(x^{j-1}, d_r^j)$$

If in step 0 and step 1(b), and (c) all the x's calculated by M^2 are also in M^1, then as previously discussed step 0 and step 1 become procedure 2. In such a case if the objective function f is quadratic, finite convergence would occur.

For the purposes of the next theorem it is imperative to notice that in steps 0 and 1 if the convex-simplex method does not first change a component of x, then that component will not be changed in those steps. To be explicit, suppose step 0 has occurred and $[x^{j-1}]_l$ is a specific value. If $[d^j_r]_l = 0$, then $[x^{j-1}]_l = [x^{j-1}_{r+2}]_l$. Suppose we continue to step 1(a) and the convex-simplex method does not change $[x^{j-1}_{r+2}]_l$, then again $[x^j_1]_l = [x^{j-1}]_l$. Should we complete step1(b), since $[d^j_i]_l = 0$, $i = 1, \ldots, r$, $[x^j_{r+1}]_l = [x^{j-1}]_l$. Then in step 1(c) because $[x^{j-1}_{r+2}]_l = [x^{j-1}]_l$

$$[d^j_{r+1}]_l = [x^{j-1}]_l - [x^{j-1}]_l = 0$$

and clearly,

$$[x^j_{r+2}]_l = [x^{j-1}]_l$$

Continuing in this manner, we find the lth component of x will remain unchanged in steps 0 and 1 unless the convex-simplex method first adjusts $[x]_l$.

To reiterate, if the convex-simplex method stops changing $[x]_l$, then after the next time step 0 occurs, $[x]_l$ will no longer be altered in steps 0 and 1.

Quadratic Convergence in a Finite Number of Steps

Now consider the QP problem. Suppose the assumptions hold which ensure that the convex-simplex method working alone converges to a point which satisfies the K-T conditions. By Convergence Theorem B the CSM-CD method will also converge to such a point, because the convex-simplex method is being utilized both in step 0 and step 1. However, the convergence might be only in a limiting sense. The following theorem ensures that the convergence is finite. A solution point is defined as a point at which the K-T conditions hold.

To simplify the proof we first impose two assumptions.

Assumption I: The CSM-CD method converges to a unique solution point.

Since we know that any convergent subsequence must converge to a solution, Assumption I guarantees that the entire sequence converges to a unique point x^*.

Assumption II: Consider any tableau T at a solution x^*. Suppose $[x^*]_l = 0$ and $[x^*]_l$ is not in the basis. Then, if $[c(x^*)]_l$ is the corresponding relative cost, $[c(x^*)]_l < 0$.

The K-T conditions ensure that $[c(x^*)]_l \leq 0$. The strict inequality on the relative cost can be considered a nondegeneracy assumption.

Both assumptions I and II nearly always hold in practice. Moreover, although we can establish the theorem under weaker hypotheses, such hypotheses complicate the mathematics and obfuscate the essential points in the proof.

Theorem 9.1: *Let p be an arbitrary quadratic function for problem* (9.1). *Suppose assumptions I and II and the assumptions of the convex-simplex method hold. Then the CSM-CD method converges in a finite number of steps.*

Note: In the theorem x^k denotes the kth point generated not the x^j in the procedure.

PROOF: Suppose the procedure does not terminate in a finite number of iterations and that it generates an infinite sequence. As mentioned above, assumption I ensures that the entire sequence must converge to x^*

$$\lim_{k \to \infty} x^k \longrightarrow x^* \tag{9.6}$$

Let γ be the number of the components of x^* that are positive. Denoting $[x]_l$ as the lth component of x, we let

$$[x^*]_l = 0 \qquad l \in L = \{\gamma + 1, \ldots, m\} \tag{9.7}$$

and

$$[x^*]_l > 0 \qquad l = 1, \ldots, \gamma$$

Then $m \leq \gamma$ because at least m variables are positive by the convex-simplex method nondegeneracy assumption.

From (9.6) for k large enough the $[x^k]_l$, $l \in L$, are very near to zero. Then because we choose the largest variables for the basis, the variables

$$[x^k]_l \qquad l \in L$$

will remain nonbasic after a finite number of iterations, say after iteration P^0.

Calling upon assumption II, we arrive at

$$[c(x^*)]_l < 0 \qquad l \in L$$

and from the continuity of $c(x)$

$$[c(x)]_l < 0 \qquad l \in L \tag{9.8}$$

for all x in a neighborhood of x^*. Hence, for k sufficiently large, say $k > P^1 \geq P^0$, by (9.6)

$$[c(x^k)]_l < 0 \qquad l \in L \tag{9.9}$$

We will now show that an integer P^3 exists such that for all $k \geq P^3$

$$[x^k]_l = 0 \qquad l \in L \tag{9.10}$$

Select a $k > P^1$ on which step 2 occurs, and suppose for some $l \in L$ that

$$[x^k]_l > 0 \tag{9.11}$$

Then by (9.11) and (9.9) step 2 must decrease $[x^k]_l$. Furthermore, for k sufficiently large by (9.6) and (9.7) the basic variables will all be large in comparison to $[x^k]_l$. Thus for k sufficiently large, when $[x^k]_l$ decreases, it will be possible to decrease $[x^k]_l$ to zero without forcing a basic variable to zero and hence without moving outside the feasible region. Moreover, the $[x^k]_l$ must be forced all the way to zero because by (9.8) the objective function will obtain its largest increase in doing so. Consequently, for some iteration k with k sufficiently large, say $k > P^2 \geq P^1$, if $[x^k]_l > 0$, $l \in L$, then step 2 will decrease $[x^k]_l$ to zero.

For all subsequent iterations k, $[x^k]_l$ will remain zero. Certainly, step 2 cannot increase $[x]_l$. Also by (9.9) the convex-simplex method will not increase $[x^k]_l$. Similarly, as discussed after presentation of the algorithm, steps 0 and 1, because they are based upon the convex-simplex method, will then never increase $[x^k]_l$. Repeating this argument for all $l \in L$, we find there must be a $P^3 \geq P^2$ such that (9.10) holds.

Now for $k > P^3$ the components $[x^k]_l$, $l = 1, \ldots, \gamma$, are strictly positive, and

$$[x^k]_l = 0 \qquad l \in L \tag{9.12}$$

Thus only the variables $[x^k]_l$, $l = 1, \ldots, \gamma$, are being adjusted, and none of them are driven to zero. Consequently, the maximum in map M^2 must be occurring before a boundary is reached and M^2 is yielding points that are in M^1. Steps 0 and 1 of the CSM-CD procedure are then behaving like procedure 2, and it follows that conjugate directions are being generated.

Consequently, the procedure must reach the solution point in a finite number of iterations after P^3. ◆

9.1.1. An Example of the CSM-CD Algorithm

Suppose we must solve the problem

$$\min \quad [x]_1^2 + [x]_2[x]_3 + [x]_4^2$$
$$\text{subject to} \quad [x]_1 + [x]_2 + [x]_3 + [x]_4 \geq 4$$
$$[x]_1 + [x]_2 \qquad\qquad \geq 2$$
$$[x]_i \geq 0$$

Note we must minimize instead of maximize; hence in calculating the relative costs, we will interchange the use of max and min. Furthermore, the objective function is not convex.

After addition of two slack variables $[x]_5$ and $[x]_6$, $m = 2$, $n = 6$, and $r = 4$.

Suppose an appropriate phase I procedure yields x^0 and the corresponding tableau as follows:

	$[x]_1$	$[x]_2$	$[x]_3$	$[x]_4$	$[x]_5$	$[x]_6$
x_0	2	0	2	0	0	0
T^0	1	1	0	0	0	−1
	0	0	1	1	−1	1

Here $[x]_1$ and $[x]_3$ are in the basis.

Observe x^0 is feasible for the original problem.

Step 0

Increase j so that $j = 1$. (To simplify notation transposes on vectors will be omitted.)

Set

$$d_1^1 = (0, 0, 0, 0, 0, 0)$$
$$d_2^1 = (0, 0, 0, 0, 0, 0)$$
$$d_3^1 = (0, 0, 0, 0, 0, 0)$$

Now employ the convex-simplex method

$$\nabla f(x^0) = (2[x^0]_1, [x^0]_3, [x^0]_2, 2[x^0]_4, 0, 0) = (4, 2, 0, 0, 0, 0)$$
$$\nabla f_B(x^0) = (2[x^0]_1, [x^0]_2) = (4, 0)$$

$$c^0 = \left(\nabla f(x^0) - (T^0)^t \nabla f(x^0)_B \right)$$

$$c^0 = \begin{pmatrix} 0 \\ -2 \\ 0 \\ 0 \\ 0 \\ 4 \end{pmatrix}$$

$$c_{s_1}^0 = \min_j \ \{c_j^0\} = -2 : s_1 = 2$$
$$c_{s_2}^0 x_{s_2}^0 = \max_j \ \{c_j^0 x_j^0\} = 0$$

Since $|c^0_{s_1}| > c^0_{s_2} x^0_{s_2}$, we increase $[x]_2$

$$\Delta = \min \ \left\{ \frac{[x^0]_{B_i}}{t^0_{i2}} \middle| t^0_{i2} > 0 \right\} = 2$$

Using

$$[y^0]_i = [x^0]_i \quad \text{for all nonbasic variables except} \quad [x]_2$$
$$[y^0]_2 = [x^0]_2 + \Delta$$

and

$$[y^0]_i = [x^0]_i - t^0_{i2}\Delta \quad \text{for all basic variables}$$

we arrive at

$$y^0 = (0, 2, 2, 0, 0, 0)$$

To obtain x^0_6, calculate

$$\begin{aligned}
f(x^0_6) &= \min_{0 \leq \tau \leq 1} \ \{f(x^0) + \tau(y^0 - x^0)\} \\
&= \min_{0 \leq \tau \leq 1} \ \{(2 - \tau 2)^2 + (\tau 2)(2 + \tau(2 - 2))\} \\
&= \min_{0 \leq \tau \leq 1} \ \{4(1 - \tau)^2 + 4\tau\}
\end{aligned}$$

But

$$0 = \frac{df}{d\tau} = -8(1 - \tau) + 4$$

and when solved

$$\tau^* = \tfrac{1}{2}$$

Therefore

$$x^0_6 = (1, 1, 2, 0, 0, 0)$$

and

$$d^1_4 = x^0_6 - x^0_1 = (-1, 1, 0, 0, 0, 0)$$

We may select $[x^0_6]_1$ and $[x^0_6]_3$ for the basis, and the tableau remains the same.

Because $\tau^* = \tfrac{1}{2}$,

$$x^0_6 \in M^1(x^0, d^1_4)$$

Therefore we go to step 1.

Step 1: So far

$$d_1^1 = d_2^1 = d_3^1 = 0$$

$$d_4^1 = (-1, 1, 0, 0, 0, 0)$$

and

x_6^0	1	1	2	0		0	0
T^0	1	1	0	0		0	1
	0	0	1	1		-1	1

(a) Using x_6^0 and the tableau T^0, we find a convex-simplex method step yields

$$x_1^1 = (1, 1, \tfrac{3}{2}, \tfrac{1}{2}, 0, 0)$$

Again no pivoting is required.

(b) Since $d_i^1 = 0$, $i = 1, 2, 3$,

$$x_i^1 = x_1^1 \qquad i = 2, 3, 4$$

Now calculate $x_5^1 \in M^2(x_4^1, d_4^1)$. Here we exploit exercise 9.2.

$$\tau_1 = \begin{cases} -\infty & \text{if } [d_4^1]_l \le 0 & \text{for all } l \\ \max \left\{ \frac{-[x_4^1]_l}{[d_4^1]_l} \Big| [d_4^1]_l > 0 \right\} & \text{otherwise} \end{cases} = \max \ \frac{-1}{1} = -1$$

$$\tau_2 = \begin{cases} +\infty & \text{if } [d_4^1]_l \ge 0 & \text{for all } l \\ \min \left\{ \frac{-[x_4^1]_l}{[d_4^1]_l} \Big| [d_4^2]_l < 0 \right\} & \text{otherwise} \end{cases} = \min \ \frac{-1}{-1} = 1$$

Thus to obtain x_5^1, calculate

$$\min_{-1 \le \tau \le 1} \{ f(x_4^1 + \tau d_4^1) \} = \min_{-1 \le \tau \le +1} \{ (1 - \tau)^2 + (\tfrac{3}{2})(1 + \tau) + (\tfrac{1}{2})^2 \}$$

Now

$$0 = \frac{df}{d\tau} = 2(1 - \tau) + \tfrac{3}{2} \quad \text{yielding} \quad \tau^* = \tfrac{1}{4}$$

Therefore,

$$x_5^1 = (\tfrac{3}{4}, \tfrac{5}{4}, \tfrac{3}{2}, \tfrac{1}{2}, 0, 0)$$

Also

$$x_5^1 \in M^1(x_4^1, d_4^1)$$

and we continue to (c).

(c) $d_5^1 = x_5^1 - x_6^0 = (-\tfrac{1}{4}, \tfrac{1}{4}, -\tfrac{1}{2}, \tfrac{1}{2}, 0, 0)$

Computing, we get

$$x_6^1 = (\tfrac{2}{3}, \tfrac{4}{3}, \tfrac{4}{3}, \tfrac{2}{3}, 0, 0)$$

Furthermore $x_6^1 \in M^1(x_5^2, d_5^1)$, and we continue to (d).

(d) $d_1^2 = (0, 0, 0, 0, 0, 0)$ $d_2^2 = (0, 0, 0, 0, 0, 0)$
          $d_3^2 = (-1, 1, 0, 0, 0, 0)$ $d_4^2 = (-\tfrac{1}{4}, \tfrac{1}{4}, -\tfrac{1}{2}, \tfrac{1}{2}, 0, 0)$

Go to step 1 with $j = 2$.

Step 1

(a) With $x_6^1 = (\tfrac{2}{3}, \tfrac{4}{3}, \tfrac{4}{3}, \tfrac{2}{3}, 0, 0)$, $[x_6^1]_2$ and $[x_6^1]_3$ are in the basis. Therefore the corresponding tableau is

x_6^1	$\tfrac{2}{3}$	$\tfrac{4}{3}$	$\tfrac{4}{3}$	$\tfrac{2}{3}$	0	0
T_6^1	1 1 0 0			0		−1
	0 0 1 1			−1		1

with $x_B = ([x]_2, [x]_3)$.

The spacer step reveals x_6^1 is a solution, with objective-function value

$$f(x_6^1) = 2\tfrac{2}{3}$$

Illustration of Step 2

The previous example, to preserve clarity and brevity, did not illustrate step 2. Suppose that with the above example on some iteration we entered step 2. Assume $j = 3$ and that the variable x^3 and tableau T^3 are of the form:

x^3	0	2	$\tfrac{5}{2}$	0	$\tfrac{1}{2}$	0
T^3	1 1 0 0			0		−1
	0 0 1 1			−1		1

Here $x_B = ([x^3]_2, [x^3]_3)$, and $\nabla f(x^3)_B = (\tfrac{5}{2}, 2)$.

Then

$$c^3 = \begin{pmatrix} -\tfrac{5}{2} \\ 0 \\ 0 \\ -2 \\ +2 \\ +\tfrac{1}{2} \end{pmatrix} \quad \text{and} \quad c^3 x^3 = \begin{pmatrix} 0 \\ 0 \\ 0 \\ 0 \\ 1 \\ 0 \end{pmatrix}$$

Set $l = 1$. Observe that, because we are minimizing the objective function, the inequalities for selection of the decreasing variable are reversed. Clearly $[c^3]_1[x^3]_1 = 0$, so $l = 2$. Then $[c^3]_2[x^3]_2 = 0$, and l increases to 3. Also $[c^3]_3[x^3]_3 = [c^3]_4[x^4]_4 = 0$. Now $l = 5$. But

$$[c^3]_5[x^3]_5 > 0$$

Therefore decrease $[x]_5$. The process of decreasing $[x]_5$ is analogous to the convex-simplex method case (c). Using the notation of case (c), we find

$$\Delta_1^3 = -\tfrac{3}{2}$$

$$\Delta_2^3 = -\tfrac{1}{2}$$

and

$$\Delta^3 = -\tfrac{1}{2}$$

Also

$$y^3 = \begin{pmatrix} 0 \\ 2 \\ 2 \\ 0 \\ 0 \\ 0 \end{pmatrix}$$

We obtain x^4 by calculating

$$\max_{0 \le \tau \le 1} \ f(x^3 + \tau(y^3 - x^3))$$

Then

$$x^4 = (0, 2, 2, 0, 0, 0)$$

Set $j = 4$. No pivoting is necessary, and

$$x_B = ([x]_2, [x]_3) = (2, 2)$$
$$\nabla f(x^4) = (0, 2, 2, 0, 0, 0) \qquad \nabla f(x^4)_B = (2, 2)$$
$$c^4 = (-2, 0, 0, -2, +2, 0) \quad \text{and} \quad l = 6$$

Clearly $[c^4]_6[x^4]_6 = 0$. Set $l = 7$. Then from step 2 (a) we return to step 0 and continue as previously described.

9.2. QUADRATIC PROGRAMMING BY MANIFOLD SUBOPTIMIZATION

We now apply the manifold-suboptimization method of Chap. 8 to the QP problem:

$$\max \quad q^t x + \tfrac{1}{2} x^t Q x$$
$$Ax = b \qquad \qquad (9.13)$$
$$x \geq 0$$

where the objective function is concave and Q is symmetric and negative semidefinite. Our approach will be to present an algorithm for problem (9.13) similar to the simplex method in that it is based upon pivoting in certain tableaux. We will then demonstrate that the pivoting algorithm behaves precisely as the manifold-suboptimization method, and hence the pivoting procedure will converge in a finite number of pivots. A knowledge of the manifold method is assumed.

A Pivot Procedure

Let the initial tableau be of the form

$$Ax \qquad \qquad = b$$
$$Qx + A^t u + \lambda = -q \qquad \qquad (9.14)$$

Observe if we determine a point (x, u, λ) that satisfies (9.14) for which

$$x \geq 0 \qquad \qquad (9.15)$$
$$\lambda \geq 0 \qquad \qquad (9.16)$$

and

$$x^t \lambda = 0 \qquad \qquad (9.17)$$

then the K-T conditions hold for problem (9.13).

Pivoting

The basic variables for the above tableau (9.14) will possess a special form in that the variables u will always be in the basis, while the variables x

and λ will pivot in and out of the basis. A nondegeneracy assumption will also be imposed requiring all basic variables to be nonzero. Since the tableaux is $(m + n) \times (2n + m)$ in size, the algorithm will always generate $n + m$ nonzero variables which will comprise the basis. By the nondegeneracy assumption any zero variable cannot be in the basis.

Complementarity and Semicomplementarity

Before stating the algorithm, we need some definitions. A point is termed complementary if it satisfies (9.14), (9.15), and

$$x_i \lambda_i = 0 \qquad i = 1, \ldots, n$$

A point is called semicomplementary if it satisfies (9.14), (9.15), and if for some j

$$x_j \neq 0 \quad \text{and} \quad \lambda_j \neq 0$$

and for some r

$$x_r = 0 \quad \text{and} \quad \lambda_r = 0$$

but for all other i

$$x_i > 0 \quad \text{implies} \quad \lambda_i = 0 \quad \text{and} \quad \lambda_i \neq 0 \quad \text{implies} \quad x_i = 0$$

Observe by the nondegeneracy assumption that complementary and semi-complementary points have precisely $n + m$ nonzero components. These components form the basis.

We can now give a precise statement of the algorithm.

A Pivot Algorithm

Initialization Step: Let (x^1, u^1, λ^1) be complementary. Go to step (a) with $k = 1$.

Step (a): The point (x^k, u^k, λ^k) is complementary. Determine

$$\lambda_j^k = \min \ \{\lambda_i^k\}$$

If $\lambda_j^k \geq 0$, terminate. The K-T conditions are satisfied, and x^k is optimal.

If $\lambda_j^k < 0$, introduce x_j into the basis by increasing it and adjusting only basic variables. If λ_j drops, i.e., becomes zero, let the corresponding $x = x^{k+1}$. Go to step (a) with $k + 1$ replacing k. If x_r drops, let the corresponding $x = x^{k+1}$, and go to step (b) with $k + 1$ replacing k.

Step (b): The point is semicomplementary $(x_j^k \neq 0, \lambda_j^k \neq 0, x_r^k = 0, \lambda_r^k = 0)$. Introduce λ_r into the basis by increasing it and adjusting only basic variables. If λ_j drops, the point is complementary. In this case go to step (a) with $k + 1$ replacing k and the corresponding $x = x^{k+1}$. If some x_{r_2} drops, repeat step (b) with r_2 playing the role of r, the corresponding $x = x^{k+1}$, and $k + 1$ replacing k.

Comments on the Algorithm

Note that at the end of step (a) or (b) the point is either complementary or semicomplementary. This is obvious in all cases except perhaps step (b) when x_{r_2} hits zero. Then $x_j \neq 0, \lambda_j \neq 0; x_{r_2} = 0, \lambda_{r_2} = 0; x_r = 0;$ and $\lambda_r \neq 0$. Also observe that throughout the procedure $x \geq 0$.

Each point (x^k, u^k, λ^k) has $n + m$ nonzero components which comprise a basis, and corresponding to each basis is a tableau. Thus, just as in the simplex method, instead of physically adjusting variables, we may pivot to generate the next tableau (exercise 9.5).

Relation of the Pivoting Algorithm to the Manifold-Suboptimization Procedure

We will now demonstrate that the pivot algorithm is a special case of the manifold-suboptimization procedure by giving each method the same point x^k and proving that the x^{k+1} generated by each is the same. The assumptions required are that the quadratic objective function be concave, that the solutions of the subproblems exist, be unique, and satisfy the uniqueness property specified in Chap. 8. These requirements ensure that the manifold method converges. As previously mentioned, we also require that all bases to (9.14) be nondegenerate.

Step (a) Occurs at x^k

To begin, assume that at x^k step (a) of the pivot method occurs so that x^k is complementary. By nondegeneracy only nonbasic x^k components are zero. Thus in the notation of the manifold method

$$B(x^k) = \{i \,|\, x_i^k \quad \text{not in the basis}\}$$

The complementarity ensures that x^k is the optimal solution to the subproblem

$$\max \ \{f(x) \,|\, x \in Y_{B(x^k)}\}$$

and that x^k is feasible for problem (9.13). In terms of the manifold method case 1 has occurred at x^k. More precisely, in the manifold terminology, y^{k-1} is feasible for problem (9.13), and we have set $x^k = y^{k-1}$. Also by nondegeneracy $B^{k-1} = B(x^k)$. We therefore see that if step (a) occurs in the pivot procedure, case 1 has occurred in the manifold method.

In both methods if all $\lambda_j \geq 0$, we terminate.

Suppose $\lambda_j < 0$. The manifold method forms

$$B^k = B^{k-1} - \{j\}$$

and solves the corresponding subproblem

$$\max \ \{f(x)\,|\,x \in Y_{B^k}\} \tag{9.18}$$

for the optimal point y^k.

In the tableau note that the subproblem (9.18) can be solved by increasing the nonbasic x_j until the basic variable $\lambda_j = 0$ (exercise 9.4). When $\lambda_j = 0$, the corresponding x is precisely the point y^k of the manifold method.

If the y^k is feasible for problem (9.13), then in the pivot procedure λ_j drops first, and in both the manifold method and the pivot algorithm $x^{k+1} = y^k$. In this situation x^{k+1} is complementary in the pivot procedure, and case 1 occurs in the manifold method. Thus the two methods generate the same point x^{k+1}.

Suppose, however, that y^k is infeasible. Then by increasing x_j in the tableau, we find some basic variable becomes zero before λ_j becomes zero. In the pivot procedure the corresponding x is x^{k+1}. Observe that by increasing x_j and adjusting only basic variables we geometrically move in a straight line toward y^k. For the manifold procedure x^{k+1} is the feasible point furthest from x^k on the straight line between x^k and y^k. But this is just the place where in the pivot procedure $x_r = 0$. Hence again in both the pivot and manifold procedure the same point x^{k+1} is generated.

Also note that in this situation x^{k+1} is semicomplementary. Moreover, in general x^{k+1} is semicomplementary whenever the y^k for the previous manifold-suboptimization problem is infeasible.

Step (b) Occurs at x^k

Now suppose a point x^k has been generated by both procedures but that step (b) has occurred at x^k in the pivot procedure. Then $x_j^k \neq 0$, $\lambda_j^k \neq 0$, $x_r^k = 0$, and $\lambda_r^k = 0$. As just mentioned, this point x^k would be the result in the manifold procedure of having y^{k-1} infeasible.

At x^k the manifold procedure solves the subproblem

$$\max\ \{f(x)\,|\,x \in Y_{B^k}\}$$

for y^k. Here, due to nondegeneracy,

$$B^k = B^{k-1} \cup \{r\}$$

where x_r is the component of x that was driven to zero when constructing x^k from x^{k-1} and the infeasible y^{k-1}. In the tableau the point y^k could be constructed from x^k by increasing λ_r until $\lambda_j = 0$ (exercise 9.4). Should λ_j drop first, the point is complementary, and y^k is feasible. Then in both procedures $x^{k+1} = y^k$.

If y^k is infeasible, then by increasing λ_r some basic x variable would become zero. In the pivot procedure the corresponding $x = x^{k+1}$. Moreover, the same x^{k+1} results in the manifold procedure because x^{k+1} would be determined as the feasible point most distant from x^k on the line between x^k and the infeasible y^k. Again, the same x^{k+1} is generated by both procedures.

Finite Convergence Because Subproblems Solved in One Step

Thus we see that if the manifold procedure were given the same initial point as the pivot procedure, both would generate the same points. Consequently, the pivot procedure converges after a finite number of pivots.

Essentially, given a quadratic objective function, and by using the pivot algorithm, we find that the manifold method solves each subproblem in one step. Thus if in the pivot procedure x^k is complementary and $\lambda_j^k < 0$, the subproblem

$$\max\ \{f(x)\,|\,x \in Y_{B^k}\}$$

where $B^k = B(x^k) - \{j\}$ can be solved in one step by increasing x_j until $\lambda_j = 0$. (Of course, certain other x components may then be negative.) Also if x^k is semicomplementary, the subproblem

$$\max\ \{f(x)\,|\,x \in Y_{B^k}\}$$

where $B^k = B^{k-1} \cup \{r\}$ can be solved in one step by increasing λ_r until $\lambda_j = 0$. Thus finite convergence is possible.

For general f, of course, the pivot procedure cannot be substituted for the manifold method, and the solution of each subproblem may not be a finite process.

9.1. Modify the CSM-CD procedure and Theorem 9.1 to permit the feasible region to be unbounded and the objective function to be unbounded from above on the feasible region.

9.2. Consider the CSM-CD method for the problem

$$\max \quad p(x) = \tfrac{1}{2}q'x + \tfrac{1}{2}x'Qx$$
$$Ax = b$$
$$x \geq 0$$

Let $x = (x_i)$ satisfy $Ax = b$, and let $d = (d_i)$ be a direction generated by the method.

(a) Prove that for all τ

$$A(x + \tau d) = b$$

(b) Show that to calculate $y \in M^2(x, d)$ it is sufficient to let

$$y = x + \tau' d$$

where τ' solves

$$\max_{\tau_1 \leq \tau \leq \tau_2} \quad p(x + \tau d)$$

and

$$\tau_1 = \begin{cases} -\infty & \text{if } d_i \leq 0 \qquad \text{for all } i \\ \max \left\{ \frac{-x_i}{d_i} \,\middle|\, d_i > 0 \right\} & \text{otherwise} \end{cases}$$

$$\tau_2 = \begin{cases} +\infty & \text{if } d_i \geq 0 \qquad \text{for all } i \\ \min \left\{ \frac{-x_i}{d_i} \,\middle|\, d_i < 0 \right\} & \text{otherwise} \end{cases}$$

9.3. Use the CSM-CD procedure to

$$\max \quad p(x) = q'x + \tfrac{1}{2}x'Qx$$
$$\text{subject to} \quad Ax = b$$
$$x \geq 0$$

where

$$q = \begin{bmatrix} 1 \\ 3 \\ 5 \end{bmatrix} \qquad Q = \begin{bmatrix} 10 & 6 & 1 \\ 6 & 4 & 0 \\ 1 & 0 & 1 \end{bmatrix}$$

$$A = [1, 2, -2] \quad \text{and} \quad b = [7]$$

9.4. Consider the pivot algorithm for the QP problem where $f(x) = q'x + \frac{1}{2}x'Qx$. Suppose the assumptions stated in the text hold.

(a) Prove that if (x^k, u^k, λ^k) is complementary, then x^k is optimal for the problem

$$\max \quad \{f(x) \,|\, x \in Y_B\}$$

where $B = \{i \,|\, x_i^k \text{ not in the basis}\}$.

(b) Suppose (x^k, u^k, λ^k) is complementary. Suppose also that $\lambda_j = \min \; \{\lambda_i\} < 0$. Let x' be the optimal solution to

$$\max \quad \{f(x) \,|\, x \in Y_{B'}\}$$

where

$$B' = B - \{j\}$$

Prove that x' may be found by increasing x_j in the tableau until $\lambda_j = 0$.

Hint: Consider the basic solution to the above problem. Show by nondegeneracy that $\lambda_j' = 0$ and $x_j' > 0$. Now decrease λ_j adjusting only basic variables, and prove that x_j must eventually go to zero. This point is x^k. Use the fact that the basic variables are uniquely determined as λ_j is varied and that there exists a point x^k at which $x_j^k = 0$. Now reverse the process by increasing x_j^k.

(c) Let (x^k, u^k, λ^k) be semicomplementary. Then

$$x_j^k \neq 0 \qquad \lambda_j^k \neq 0 \qquad x_r^k = 0 \qquad \lambda_r^k = 0$$

Show by increasing λ_r and adjusting only basic variables that we can find a point y' optimal to the problem

$$\max \quad \{f(x) \,|\, x \in Y_B\}$$

where

$$B = \{i \,|\, x_i^k = 0\}$$

Hint: Let (x', u', λ') be the basic solution corresponding to the optimal y'. First prove $\lambda'_r > 0$. Recall $x^k_r = 0$ because y^{k-1} is infeasible. Therefore if we decrease x_r and maintain optimality, f should increase. Then from Sec. 3.2 $\lambda'_r \geq 0$. But nondegeneracy implies $\lambda'_r \neq 0$, therefore $\lambda'_r > 0$. Now reason as in part (b) above.

9.5. Indicate how to determine the pivot element for updating the tableau in the pivot procedure for the QP problem. Consider all cases.

9.6. Develop a procedure for determining a complementary solution for the pivot algorithm. This point would serve as an initial point for the algorithm.

9.7. Consider the QP problem

$$\begin{aligned}
\min \quad & q^t x + \tfrac{1}{2} x^t Q x \\
& 3x_1 + 2x_2 + x_3 = 6 \\
& x_1 \geq 0 \qquad x_2 \geq 0 \qquad x_3 \geq 0
\end{aligned}$$

where

$$q = \begin{pmatrix} 1 \\ 2 \\ 0 \end{pmatrix} \quad \text{and} \quad Q = \begin{bmatrix} 25 & 4 & 10 \\ 4 & 1 & 1 \\ 10 & 1 & 21 \end{bmatrix}$$

Solve by the pivot algorithm. Show that the manifold-suboptimization algorithm, if it starts at a point acceptable to the pivot algorithm, generates the same points as the pivot algorithm.

NOTES AND REFERENCES

§9.1. This section is new. Computer tests indicate that the CSM-CD procedure is quite efficient for general f. For related applications of conjugate directions see FAURE and HUARD; ZOUTENDIJK (1960).

§9.2. The pivot algorithm was suggested in papers by Dantzig [DANTZIG (1963)] and by VAN DE PANNE and WINSTON. These papers proved convergence by tracing the algebraic manipulation of the tableau. The argument in this section is geometrical in content and relates the pivot algorithm to the more general approach of the manifold method.

ADDITIONAL COMMENTS

Quadratic programming has been studied in detail. Usually it is treated as an extension of LP; this text, however, treats it as a specialization of NLP. The NLP approach permits, for example, the CSM-CD method to be applicable to general functions as well as to quadratic functions. For other work on QP, see BEALE (1959); BARANKIN and DORFMAN; BOOT (1964); HILDRETH (1957); HOUTHAKKER; LEMKE (1962); MARKOWITZ (1956); THEIL and VAN DE PANNE; and WOLFE (1959). The problem of finding global optima for QP is treated in articles by RITTER (1965) and (1966). Parametric and sensitivity studies are examined in BOOT (1963).

A somewhat different approach to QP, suggested by Lemke, solves the K-T conditions directly via a complementary pivot algorithm. This approach has interesting applications to games; see LEMKE; LEMKE and HOWSON; and COTTLE and DANTZIG.

10 Some Procedures for the Nonlinear Programming Problem with an Application to the Liapunov Theory of Difference Equations

The general NLP problem has been defined as

$$\max \quad f(x)$$
$$\text{subject to} \quad g_i(x) \geq 0 \qquad i = 1, \ldots, m \tag{10.1}$$

and in this chapter we develop algorithms for it. The first algorithm, called the method of centers, provides a straightforward application of Convergence Theorem A by transforming the constrained problem, problem (10.1), into a sequence of unconstrained problems. The second portion of this chapter delves into a Lagrangean method for problem (10.1), which is based upon difference equations and has an interesting interpretation in terms of certain dynamic economic processes. In the final section we analyze more general dynamic processes that are modeled by difference equations. In this context Convergence Theorem A becomes synonymous with Liapunov stability theory.

10.1. THE METHOD OF CENTERS

The method of centers is in a class of methods that solve problem (10.1) by solving a sequence of unconstrained problems with the essential goal of suppressing explicit consideration of the constraints. The algorithmic map is quite simple to state. First define a function

$$G(w, x) = \min \; [f(w) - f(x), g_1(w), \ldots, g_m(w)]$$

Observe

$$G(w, x) > 0$$

if and only if

$$f(w) > f(x) \quad \text{and} \tag{10.2}$$
$$g_i(w) > 0 \quad \text{all } i$$

Then the algorithmic map A becomes

$$A(x) = \{y \mid G(y, x) = \max_{w \in E^n} \; G(w, x)\}$$

Verbally, given x^k, the successor point x^{k+1} is any point that yields the unconstrained maximum of $G(w, x^k)$. We assume that all maxima exist (exercise 10.4) and that an initial feasible point x^1 is given. Observe that then all points x^k must be feasible.

Convergence Proof

Convergence Theorem A will be employed to establish convergence. All functions f and g_i are assumed continuous, and it must also be assumed that if x^* is optimal for problem (10.1) then in any neighborhood of x^* there are points w such that

$$g_i(w) > 0 \quad \text{for all} \quad i = 1, \ldots, m$$

This assumption implies that near x^* there are feasible points w that are not on the boundary of the feasible region. Thus even if x^* is on the boundary, there are points w near x^* that are on the inside of the feasible region. As usual we assume all points x^k generated belong to a compact set so that condition 1 holds (exercise 10.4).

Condition 2

A solution is any point x^* which is optimal for problem (10.1). Also let $z = x$ and $Z(z) = f(x)$. Now consider condition 2(b); if x^* is optimal, then there is no feasible point w that yields $f(w) > f(x)$. Hence by (10.2)

$$\max_{w} \quad G(w, x^*) = G(x^*, x^*) = 0$$

Consequently

$$y \in A(x^*) \quad \text{implies} \quad f(y) = f(x^*)$$

and condition 2(b) holds.

For condition 2(a) let x^k be nonoptimal. As noted above, x^k is feasible. Then $f(x^*) > f(x^k)$, and because f is continuous we may select a point w' so close to x^* that

$$f(w') > f(x^k)$$

Also by assumption, w' may be selected so that

$$g_i(w') > 0 \quad i = 1, \ldots, m$$

Then for that w'

$$G(w', x^k) > 0 \tag{10.3}$$

and

$$\max_{w} \quad G(w, x^k) \geq G(w', x^k) > 0$$

The conclusion follows from (10.2) that

$$y \in A(x^k) \quad \text{implies} \quad f(y) > f(x^k)$$

Condition 3

Only condition 3 remains to be verified. Let

$$x^k \longrightarrow x^\infty \qquad k \in \mathscr{K}$$

and

$$x^{k+1} \longrightarrow x^{\infty+1} \qquad k \in \mathscr{K}$$

By definition

$$G(x^{k+1}, x^k) \geq G(w, x^k) \tag{10.4}$$

for $w \in E^n$.

Taking limits of (10.4)

$$G(x^{\infty+1}, x^\infty) \geq G(w, x^\infty) \tag{10.5}$$

for any $w \in E^n$. But then $x^{\infty+1}$ maximizes $G(w, x^\infty)$ over all w, and hence

$$x^{\infty+1} \in A(x^\infty)$$

(For an alternative proof of closedness see exercise 10.5.) The method of centers thus converges.

Observe that the successive points x^k are geometrically in the interior, or "center," of the feasible region because

$$g_i(x^k) > 0 \qquad i = 1, \ldots, m$$

Hence the name of the algorithm.

10.2. A LAGRANGEAN ALGORITHM

In all of the previous methods discussed, the Z function, that function which indicates the adaption and improvement of the process, was the objective function itself. However many other forms of Z are possible. In particular the Lagrangean method utilizes a Z function that is the distance between the point z^k and a saddle point. The method converges for problem (10.1) with f and g_i concave and continuously differentiable.

Saddle-Point Assumptions

As shown in Theorem 2.18 if the Lagrangean

$$L(x, \lambda) = f(x) + \lambda^t g(x)$$

possesses a saddle point $(\bar{x}, \bar{\lambda})$, then $\bar{x}$ solves problem (10.1). The Lagrangean method assumes that such a saddle point exists. Explicitly,

$$L(x, \bar{\lambda}) \leq L(\bar{x}, \bar{\lambda}) \leq L(\bar{x}, \lambda) \tag{10.6}$$

for all $x \in E^n$ and $\lambda \geq 0$.

Furthermore, it is supposed that there is a unique point $\bar{x}$ which solves (10.6). Then

$$L(x, \bar{\lambda}) < L(\bar{x}, \bar{\lambda}) \quad \text{if} \quad x \neq \bar{x} \tag{10.7}$$

(A simple condition that quarantees this assumption is provided in exercise 10.8.)

If we define the set

$$U = \{\bar{\lambda} \mid (\bar{x}, \bar{\lambda}) \text{ is a saddle point of } L(x, \lambda)\}$$

then U is the set of all $\bar{\lambda}$ for which $(\bar{x}, \bar{\lambda})$ is a saddle point. By assumption U is not empty, however, we also assume that U is compact. Although the set U usually is compact, the next lemma provides a condition that ensures its compactness.

Lemma 10.1: *Suppose there exists a point x' for which*

$$g_i(x') > 0 \qquad i = 1, \ldots, m$$

then U is compact.

PROOF: By definition of saddle point, for $\bar{\lambda} \in U$

$$f(x') + \sum_{i=1}^{m} \bar{\lambda}_i g_i(x') \leq L(\bar{x}, \bar{\lambda})$$

As $\bar{\lambda}_i \geq 0$ and $g_i(x') > 0$,

$$\bar{\lambda}_j g_j(x') \le \sum_{i=1}^{m} \bar{\lambda}_i g_i(x') \le L(\bar{x}, \bar{\lambda}) - f(x')$$

But because $g_j(x') > 0$,

$$0 \le \bar{\lambda}_j \le \frac{L(\bar{x}, \bar{\lambda}) - f(x')}{g_j(x')}$$

and each component of $\bar{\lambda}$ is bounded.

We now prove closedness of the set U. Let $\bar{\lambda}^k \longrightarrow \bar{\lambda}^\infty$ where $\bar{\lambda}^k \in U$. It must be established that $\bar{\lambda}^\infty \in U$. By definition of saddle point, for any $x \in E^n$ and $\lambda \ge 0$,

$$f(x) + \bar{\lambda}^k g(x) \le f(\bar{x}) + \bar{\lambda}^k g(\bar{x}) \le f(\bar{x}) + \lambda g(\bar{x}) \tag{10.8}$$

But taking limits of (10.8),

$$f(x) + \bar{\lambda}^\infty g(x) \le f(\bar{x}) + \bar{\lambda}^\infty g(\bar{x}) \le f(\bar{x}) + \lambda g(\bar{x})$$

and $\bar{\lambda}^\infty \in U$.

Therefore, U is closed and bounded.   ◆

The Algorithmic Map

We now define the algorithmic map. Let

$$L_x(x, \lambda) = \nabla_x f(x) + \sum_{i=1}^{m} \lambda_i \nabla_x g_i(x) \tag{10.9}$$

be the gradient of L with respect to x, and

$$L_\lambda(x, \lambda) = g(x) \tag{10.10}$$

be the gradient of L with respect to λ. Then given (x^k, λ^k) the algorithmic map yields (x^{k+1}, λ^{k+1}) using the equations

$$x^{k+1} = x^k + \tau L_x(x^k, \lambda^k) \tag{10.11}$$

$$\lambda^{k+1} = \max \; [0, \lambda^k - \tau L_x(x^k, \lambda^k)] \tag{10.12}$$

where τ, the step-length parameter, is a fixed positive scalar to be specified subsequently. The maximum operation in (10.12) is component by component. Thus (10.12) implies

$$\lambda_i^{k+1} = \max \quad [0, \lambda_i^k - \tau g_i(x^k)]$$

Observe that both x and λ are required to specify the algorithmic map. Consequently $z = (x, \lambda)$, and the algorithmic map A is a function A: $E^{n+m} \longrightarrow E^{n+m}$.

The Z Function

We define the Z function by

$$Z(z) = \min_{\bar{\lambda} \in U} \quad \{\|x - \bar{x}\|^2 + \|\lambda - \bar{\lambda}\|^2\} \tag{10.13}$$

Intuitively Z determines the distance between $z = (x, \lambda)$ and the nearest saddle point $(\bar{x}, \bar{\lambda})$. Note that from Theorem 7.1 as U is compact the minimum is achieved, while from Theorem 7.2 $Z(z)$ is continuous (see exercise 10.6). Because Z is measuring the distance, the sets

$$\{z \mid \gamma \le Z(z) \le \beta\} \quad \text{and} \quad \{z \mid Z \le \beta\} \tag{10.14}$$

are compact for any positive γ and β (see exercise 10.7).

A Solution

A point z is termed a solution if given $\epsilon > 0$

$$\|x - \bar{x}\| \le \epsilon \tag{10.15}$$

Thus z is a solution if the distance between x and the optimal point $\bar{x}$ is not greater than ϵ.

If a point z^k is a solution the algorithm terminates.

Specification of τ in the Algorithmic Map

With the solution point and Z function introduced, we can define the parameter τ in the algorithmic map.

Given an arbitrary initial point $z^1 = (x^1, \lambda^1)$ where $x^1 \in E^n$ and $\lambda^1 \ge 0$, let

$$P = \max \quad [\epsilon, Z(x^1, \lambda^1)]$$

Then we define τ by

$$\tau = \min \left\{ \frac{(\bar{x} - x)L_x(x, \lambda) - (\bar{\lambda} - \lambda)L_{\lambda}(x, \lambda)}{\|L_x(x, \lambda)\|^2 + \|L_{\lambda}(x, \lambda)\|^2} \,\middle|\, \epsilon \leq \|x - \bar{x}\|, \right.$$

$$\left. \epsilon \leq Z(z) \leq P, \bar{\lambda} \in U \right\} \tag{10.16}$$

Via (10.14) the sets over which the respective minimizations are taking place are compact, where here observe that $\epsilon \leq \|x - \bar{x}\|$ ensures

$$\epsilon \leq Z(z) \tag{10.17}$$

From the continuous differentiability of f and the g_i, the function in (10.16) is continuous. It is also positive at each point as the next lemma will indicate. We then know by Corollary 2.1.1 that

$$\tau > 0 \tag{10.18}$$

Lemma 10.2: *Let f and g_i, $i = 1, \ldots, m$, be concave and continuously differentiable.*
Then $(\bar{x} - x)^t L_x(x, \lambda) - (\bar{\lambda} - \lambda)^t L_{\lambda}(x, \lambda) > 0$ for $x \neq \bar{x}$.

PROOF: By concavity of $L(x, \lambda)$ in x

$$L(\bar{x}, \lambda) \leq L(x, \lambda) + (\bar{x} - x)^t L_x(x, \lambda) \tag{10.19}$$

From (10.10) $L_{\lambda}(x, \lambda) = g(x)$, and by definition of the Lagrangean,

$$L(x, \bar{\lambda}) - (\bar{\lambda} - \lambda)^t L_{\lambda}(x, \lambda) = L(x, \lambda) \tag{10.20}$$

But then equations (10.19) and (10.20) combine to yield

$$L(\bar{x}, \lambda) - L(x, \bar{\lambda}) \leq (\bar{x} - x)^t L_x(x, \lambda) - (\bar{\lambda} - \lambda)^t L_{\lambda}(x, \lambda) \tag{10.21}$$

By the definition of saddle point and (10.7), if $x \neq \bar{x}$

$$L(x, \bar{\lambda}) < L(\bar{x}, \bar{\lambda}) \leq L(\bar{x}, \lambda) \tag{10.22}$$

Consequently, from (10.21) and (10.22)

$$0 < L(\bar{x}, \lambda) - L(x, \bar{\lambda}) \leq (\bar{x} - x)^t L_x(x, \lambda) - (\bar{\lambda} - \lambda)^t L_{\lambda}(x, \lambda) \quad \blacklozenge$$

Convergence Proof

We now establish convergence of the Lagrangean procedure using Convergence Theorem A. For completeness we shall summarize all the assumptions required for convergence. The f and the g_i, $i = 1, \ldots, m$, are concave and continuously differentiable. A saddle point $(\bar{x}, \bar{\lambda})$ exists, and $\bar{x}$ is unique. Also the set U is compact. Convergence will now be verified.

Condition 3

Condition 3 is immediate because the algorithmic map is a continuous function.

Condition 2

To prove condition 2 some preliminary observations are required.
In the following let $L_x(x^k, \lambda^k) = L_x^k$ and $L_\lambda(x^k, \lambda^k) = L_\lambda^k$, then squaring Equation (10.12),

$$\| \lambda^{k+1} \|^2 \leq \| \lambda^k \|^2 - 2\tau(\lambda^k)'L_\lambda^k + \tau^2 \| L_\lambda^k \|^2 \qquad (10.23)$$

Also from (10.12), because $\bar{\lambda} \geq 0$,

$$-2(\bar{\lambda})'\lambda^{k+1} \leq -2(\bar{\lambda})'\lambda^k + 2\tau(\bar{\lambda})'L_\lambda^k \qquad (10.24)$$

Then using (10.23) and (10.24) and a little manipulation, we arrive at

$$\| \lambda^{k+1} - \bar{\lambda} \|^2 \leq \| \lambda^k - \bar{\lambda} \|^2 + 2\tau(\bar{\lambda} - \lambda^k)'L_\lambda^k + \tau^2 \| L_\lambda^k \|^2 \qquad (10.25)$$

In precisely the same manner as (10.25) was obtained from (10.12) we may determine from (10.11) that

$$\| x^{k+1} - \bar{x} \|^2 \leq \| x^k - \bar{x} \|^2 - 2\tau(\bar{x} - x^k)'L_x^k + \tau^2 \| L_x^k \|^2 \qquad (10.26)$$

Adding (10.25) and (10.26)

$$[\| x^{k+1} - \bar{x} \|^2 + \| \lambda^{k+1} - \bar{\lambda} \|^2] \leq [\| x^k - \bar{x} \|^2 + \| \lambda^k - \bar{\lambda} \|^2]$$
$$-\tau\{2[(\bar{x} - x^k)'L_x^k - (\bar{\lambda} - \lambda^k)'L_\lambda^k] - \tau[\| L_x^k \|^2 + \| L_\lambda^k \|^2]\} \qquad (10.27)$$

Now we may establish condition 2. Note there is no loss in generality by reversing inequalities and proving $Z(z^{k+1}) < Z(z^k)$. Suppose z^k is not optimal. Examining (10.27) and the definition Z, to prove $Z(z^{k+1}) < Z(z^k)$ we need only verify

$$2[(\bar{x} - x^k)'L_x^k - (\bar{\lambda} - \lambda^k)'L_\lambda^k] - \tau[\| L_x^k \|^2 + \| L_\lambda^k \|^2] > 0 \qquad (10.28)$$

Because z^k is not optimal and by (10.17) we have that

$$\epsilon < \| x^k - \bar{x} \| \quad \text{and} \quad \epsilon < Z(z^k) \leq P \tag{10.29}$$

Therefore using the definition of τ the left-hand side of (10.28) is larger than

$$2[(\bar{x} - x^k)^t L_x^k - (\bar{\lambda} - \lambda^k)^t L_\lambda^k] - \left(\frac{(\bar{x} - x^k)^t L_x^k - (\bar{\lambda} - \lambda^k)^t L_\lambda^k}{\| L_x^k \|^2 + \| L_\lambda^k \|^2} \right) \tag{10.30}$$

$$\times [\| L_x^k \|^2 + \| L_x^k \|^2] = [(\bar{x} - x^k)^t L_x^k - (\bar{\lambda} - \lambda^k)^t L_\lambda^k] > 0$$

where the final inequality holds by Lemma 10.2. Thus (10.28) is verified, and condition 2(a) is proved (see exercise 10.12).

Condition 2(b) holds immediately as the algorithm terminates if z^k is a solution.

Condition 2 is established.

Condition 1

Note also that condition 2 provides

$$Z(z^k) \leq Z(z^1)$$

Then by (10.14) all z^k are on a compact set and condition 1 is also established.

The algorithm converges.

Comments

We should note that the definition of τ, the step length, begs the question. The parameter τ is defined in terms of the solution; yet the purpose of an algorithm is to determine a solution. The proof should then be phrased in the following manner. Given $\epsilon > 0$, there exists a τ for which the algorithm converges. Precise calculation of τ is impossible. During actual calculation of the algorithm, τ is varied in an effort to achieve a "good" τ. In a similar vein, knowing when z is a solution may not be an easy task.

In most previous algorithms the difficult part of the proof was the closedness condition, condition 3. Usually condition 2 was straightforward. For the Lagrangean algorithm the opposite was true. Condition 3 was immediate while 2 was arduous to verify. One reason for this difference is that the Lagrangean method seeks a saddle point; thus we are solving a maximum-minimum problem. On the other hand, most other algorithms solve a pure maximum problem, and the objective function, because it always increases, can be used as the Z function. For the Lagrangean method the immediate function to be considered for the Z function is the Lagrangean itself. But, due to the maximum-minimum operation, the Lagrangean both increases and decreases and it cannot be utilized for the Z function. Instead, a complicated Z function based upon the distance to a saddle point had to be employed. Verifying condition 2, which requires that we get closer and closer to a saddle point, was then extremely difficult.

It might also be noted that we proved condition 1, the compactness condition, for this algorithm. In most previous procedures condition 1 was not established but was assumed to hold. This is because in the previous procedures postulating condition 1 is geometrically reasonable. Momentary reflection reveals that such algorithms, except for unusual situations, ought to generate points in a compact set. Regrettably this geometric insight is not so easily obtained for the Lagrangean method. Hence, a mathematical proof of condition 1 was provided. We must remark, however, that situations in which compactness does not hold are discussed in Chap. 11.

10.3. AN ECONOMIC INTERPRETATION

In Chap. 3 a saddle point of the Lagrangean was discussed in relation to a competitive-equilibrium situation where an industry wished to maximize L by selecting the operating level x, while the market wanted to minimize L through its choice of the prices λ.

The Lagrangean algorithm can be considered as a dynamic bargaining process between the industry and the market. To begin industry states its initial operating level x^1, while the market presents its initial prices λ^1. At the next round of bargaining the industry, in an effort to maximize L over x, changes x^1 in the direction of the gradient

$$x^2 = x^1 + \tau L_x(x^1, \lambda^1) \tag{10.31}$$

Similarly, the market revises λ to minimize $L(x, \lambda)$ over $\lambda \geq 0$ by moving in the direction of the negative gradient with respect to λ, unless a price is driven to zero.

$$\lambda_i^2 = \max \ [0, (\lambda_i^1 - \tau g_i(x^1))] \qquad i = 1, \ldots, m \tag{10.32}$$

Thus each side separately adjusts its offer in an effort to achieve its most desirable position. As proved, in each round of the bargaining process the two parties get closer and closer to equilibrium.

10.4. STABILITY IN DIFFERENCE-EQUATION
SYSTEMS: LIAPUNOV THEORY

The algorithmic map in the Lagrangean algorithm was actually a difference equation. In this section we extend the previous section by considering the map $A: E^n \rightarrow E^n$ to represent a general vector system of difference equations

$$z^{k+1} = A(z^k) \tag{10.33}$$

Frequently, difference equations represent the dynamic behavior of a system such as the competitive bargaining system previously described. In general, however, z^k will represent the state of the system at time k, while the function A will define the evolution of the system from one time period to the next.

Equilibrium

Of great interest are points z^* such that

$$z^* = A(z^*)$$

The difference-equation system is said to be in *equilibrium* when it is in state z^*. Clearly, once a system arrives in state z^* it remains in that state.

For most difference-equation systems we can transform coordinates such that

$$0 = A(0) \tag{10.34}$$

The point 0 will then be an equilibrium point and we term it a solution.

Asymptotic Stability

A difference-equation system represented by A is called **asymptotically stable in the large** if, given any initial point z^1,

$$\lim_{k \to \infty} z^k = 0 \tag{10.35}$$

where

$$z^{k+1} = A(z^k)$$

Asymptotic stability in the large means that, given any initial state of the system, as time progresses the system eventually evolves to the equilibrium position, state 0.

It is of great interest to know when a system is asymptotically stable in the large, and we will find the following lemma valuable in deriving stability criteria for such systems.

Lemma 10.3: *Let Z be a continuous function $Z: E^n \to E^1$. Assume*

$$Z(z) \longrightarrow +\infty \quad \text{as} \quad \|z\| \longrightarrow +\infty$$

Then for any positive number P, the set Q where

$$Q = \{z \,|\, Z(z) \leq P\}$$

is a compact set.

PROOF: By continuity of Z, Q is a closed set (see exercise 10.10). Boundedness remains to be verified.
Let

$$\gamma = \sup \{\|z\| \,|\, z \in Q\}$$

Then $\gamma < +\infty$. For if $\gamma = +\infty$, we could determine a subsequence of points $\{z^k\}_{\mathscr{K}}$ in Q such that $\|z^k\| \to +\infty, k \in \mathscr{K}$. But by hypothesis it would follow that

$$Z(z^k) \longrightarrow +\infty$$

As $Z(z^k) \leq P$ for all k, this is impossible. Hence, $\|z\| \leq \gamma < \infty$ for $z \in Q$, and Q is bounded.　◆

The main stability theorem will now be proved.

Theorem 10.4: *Consider a system of difference equations represented by a continuous function $A: E^n \to E^n$ where given z^1*

$$z^{k+1} = A(z^k) \tag{10.36}$$

and

$$0 = A(0) \tag{10.37}$$

Suppose there is a continuous function $Z: E^n \to E^1$ such that

$$Z(z) \longrightarrow +\infty \quad \text{as} \quad \|z\| \longrightarrow +\infty$$

and if $z \neq 0$

$$Z(A(z)) < Z(z) \tag{10.38}$$

Then $\lim\limits_{k \to \infty} z^k = 0$, i.e., the system is asymptotically stable in the large.

PROOF: We first establish that Convergence Theorem A holds. Exploiting (10.36), (10.37), and (10.38),

$$Z(z^{k+1}) \leq Z(z^k) \quad \text{for all} \quad k \tag{10.39}$$

From Lemma 10.3 the set

$$\{z \mid Z(z) \leq Z(z^1)\}$$

is compact. But then condition 1 of Convergence Theorem A is proved, for, using (10.39), we find all z^k are in a compact set.

Define the point 0 as the solution point. If z is not a solution, (10.38) ensures that condition 2(a) of Convergence Theorem A holds. If z is a solution, since $A(0) = 0$,

$$Z(A(0)) = Z(0)$$

and 2(b) of Convergence Theorem A is verified.

Finally condition 3 of Convergence Theorem A holds immediately because A is a continuous function.

Thus the hypotheses of Convergence Theorem A are established, and it follows that for any convergent subsequence $\{z^k\}_{\mathscr{K}}$

$$z^k \longrightarrow 0 \qquad k \in \mathscr{K}$$

But as all z^k are on a compact set and all convergent subsequences have zero as a limit point,

$$\lim_{k \to \infty} z^k = 0 \quad \blacklozenge$$

Comment

This theorem connects the theory of algorithmic convergence with the stability theory of difference or differential equations. Such stability theory is generally considered under the heading of Liapunov stability theory, and a Z satisfying the hypotheses of Theorem 10.4 would be referred to as a Liapunov function.

In physical systems the function Z can often be represented by the kinetic energy of the system. Over time, the energy of the system dissipates, leaving the system in a stable-equilibrium state.

Some applications of the theorem are given in exercise 10.9, and an insight into the Liapunov function is provided by exercise 10.11.

EXERCISES

10.1. Prove convergence of the method-of-centers procedure by using $G^1(w, x)$ instead of $G(w, x)$ where

$$G^1(w, x) = (f(w) - f(x)) \prod_{i=1}^{m} g_i(w)$$

Here $\prod\limits_{i=1}^{m}$ denotes the product.

10.2. The method of centers in the text assumed that all constraints were placed into the $G(w, x)$ function. Instead, suppose certain constraints, say the constraints $g_l, g_{l+1}, \ldots, g_m$, were handled explicitly.

Define

$$G^2(w, x) = \min \quad [f(w) - f(x), g_1(w), \ldots, g_{l-1}(w)]$$

and let the algorithmic map A be the set of all y for which

$$G^2(y, x) = \max \quad \{G^2(w, x) \,|\, g_l(w) \geq 0, g_{l+1}(w) \geq 0, \ldots, g_m(w) \geq 0\}$$

where y is also required to satisfy

$$g_i(y) \geq 0 \qquad i = l, l+1, \ldots, m$$

Intuitively, given x^k, the successor point x^{k+1} is any point y that solves the problem

$$\max \quad \{G^2(w, x^k) \,|\, g_i(w) \geq 0 \qquad i = l, l+1, \ldots, m\}$$

Study the convergence of this procedure. When would this approach be easier to use than the method of centers described previously?

10.3. How would you adapt the unconstrained methods of Chap. 5 to maximize the $G(w, x^k)$ function in the method of centers.

10.4. For the method of centers suppose that the feasible region F is compact and that $G(w, x)$ is continuous.

    (a) Prove a maximizing point exists when applying the map A.

    (b) Prove all points x^k generated are in a compact set.

10.5. (VEINOTT) For the method of centers suppose F is compact and that $G(w, x)$ is continuous.

    (a) Show that given $x \in F$

$$\max_{w \in E^n} \ G(w, x) = \max_{w \in F} \ G(w, x)$$

    (b) Show that the algorithmic map may be defined by

$$A(x) = \{y \in F \,|\, G(y, x) = \max_{w \in F} \ G(w, x)\}$$

    (c) Prove this algorithmic map is closed via Lemma 7.3.

10.6. Establish that in the Lagrangean algorithm

$$Z(z) = \min_{\bar{\lambda} \in U} \ \{\|x - \bar{x}\|^2 + \|\lambda - \bar{\lambda}\|^2\}$$

is a continuous function by application of Theorem 7.2.

10.7. Show that the sets

$$\{z \,|\, a \le Z(z) \le \beta\} \quad \text{and} \quad \{z \,|\, Z(z) \le \beta\}$$

are compact for any $0 \le a \le \beta$.

    Hint: The set U is itself compact, and $Z(z) \to +\infty$ as $\|z\| \to +\infty$ (see also Lemma 10.3).

10.8. Prove that if f is strictly concave (see exercise 2.9) then there is a unique $\bar{x}$ that yields the saddle point in (10.6). Also verify Equation (10.7).

10.9. Study the stability of the following difference-equation systems

    (a) $x_1^{k+1} = \frac{1}{3}x_1^k + \frac{1}{4}x_2^k$

        $x_2^{k+1} = \frac{1}{10}x_1^k + \frac{1}{2}x_3^k$

        $x_3^{k+1} = \frac{1}{2}x_1^k$

Hint: $Z(x) = \sum_{i=1}^{3} |x_i|.$

    (b) $x_1^{k+1} = x_1^k \dfrac{((x_1^k)^2 + (x_2^k)^2)}{((x_1^k)^2 + (x_2^k)^2 + 1)}$

        $x_2^{k+1} = x_2^k \dfrac{((x_1^k)^2 + (x_2^k)^2)}{((x_1^k)^2 + (x_2^k)^2 + 1)}$

Hint: $Z(x) = \|x\|.$

10.10. Let $h: E^n \longrightarrow E^1$ be a continuous function. Prove the set

$$\{x \,|\, h(x) \leq P\}$$

is a closed set.

10.11. Suppose that there exists a Liapunov function Z which satisfies the hypotheses of Theorem 10.4. Prove that Z is positive definite, that is

$$Z(z) > 0 \qquad z \neq 0$$
$$Z(0) = 0$$

10.12. Equation (10.29) states that $Z(z^k) \leq P$ for all k. Yet Equation (10.17) only verifies this for $k = 1$. Prove (10.29) holds for all k.

Hint: Prove it for $k = 2$, and then use induction.

NOTES AND REFERENCES

§10.1. The algorithm was developed by HUARD. The proof, of course, being based on the Convergence Theorem A, is new. For the method of centers in more abstract spaces, refer to BUI-TRONG-LIEU and HUARD.

§10.2. The Lagrangean algorithm is an adaptation of Uzawa's difference equation approach to the differential-equation technique of Arrow and Hurwicz. However it is somewhat different than Uzawa's method due to an error in the originator's proof. See ARROW, HURWICZ, UZAWA (1958); ARROW (1951); and ARROW and HURWICZ (1950).

An extension of this procedure to n-person games may be found in ROSEN (1965).

The fact that the Z function was so difficult to construct provides insight into an interesting approach suggested by WILSON. This method attempts to find the saddle point by successively taking quadratic approximations. However, because there is no Z function, the method may not adapt and cannot guarantee to converge. Indeed, experience has revealed that it often does not converge. It is quite rapid, however, when it does converge. In practice, several initial points are tried until one is found from which the procedure converges.

§10.3. For other economic interpretations refer to HURWICZ; ARROW and HURWICZ (1960).

§10.4. Liapunov theory is extremely valuable and interesting. Investigate, for example, the works of LIAPUNOV; LA SALLE and LEFSCHETZ; and KALMAN and BERTRAM.

ADDITIONAL COMMENTS

Numerous other procedures for the NLP problem have been suggested. The paper by BREGMAN discusses how to find feasible points of convex sets and relates this to linear and quadratic programming. Other procedures include those found in EREMIN; ROSEN (1961); WARGA (1963a), (1963b); WEGNER; and ZUHOVICKII, POLJAK, and PRIMAK (1963).

Although the topic of decomposing NLP problems is quite difficult, some preliminary results may be of interest. See, for example, ROSEN (1963); ROSEN and ORNEA; SANDERS; and ZANGWILL (1967h). Decomposition attempts to solve a large problem by solving several smaller problems.

Another topic that is beyond the scope of this text is parametric NLP in which we are interested in how the optimal solution varies as certain parameters change. Two papers on this topic are GEOFFRION (1966), (1967).

11 Necessary and Sufficient Conditions for Convergence

The previously developed Convergence Theorems A and B are powerful tools and, as has been demonstrated, provide a straightforward approach to proving algorithmic convergence. Nevertheless, certain algorithms, such as those presented in the next two chapters, converge yet do not satisfy the hypotheses of these theorems. In this chapter more general convergence theorems will be stated to handle such algorithms. These new theorems provide conditions that not only are sufficient for convergence but also are necessary for convergence. Specifically, a very general definition of algorithmic convergence will be stated, and then a theorem will specify conditions that hold if and only if the algorithm converges. Other, related theorems will also be developed.

11.1. INTRODUCTION

To develop these new conditions let us examine the previous convergence conditions in detail. Four difficulties are immediately evident. The first is the requirement that the algorithmic map be a closed map; many are not closed. The second difficulty is the requirement of strict adaption or improvement of the Z function if a point z^k is not a solution, condition 2(a) of Convergence Theorem A. Certainly, algorithms need not strictly adapt each time. In fact, it might be reasonable to spend a few iterations searching for the "best" move, and adaption would then occur only after a finite number of iterations L. Indeed, when adaption does occur, strict improvement $Z(z^{k+1}) > Z(z^k)$ with the strict inequality may not occur; rather the improvement may be $Z(z^{k+1}) \geq Z(z^k)$.

The compactness assumption is a third limitation of the previous convergence theorems. However, as noted in Chap. 7 some functions do not attain their maxima due to lack of compactness. Therefore, some type of assumption about compactness is required, although the new assumption will be considerably weaker than those previously stated.

The final problem regards the actual definition of the algorithmic map. We previously specified that $A_k: V \longrightarrow V$, that is, the map A_k was defined on all of the space V. Certainly A_k need only be defined on a portion of V, say $V_k \subset V$; then $A_k: V_k \longrightarrow V_{k+1}$.

To begin our detailed examination of these four difficulties let us incorporate the revised concept of algorithmic map $A_k\colon V_k \longrightarrow V_{k+1}$, into a new definition of algorithm. In the new definition, we employ set notation to define termination of the algorithm. If

$$A_k(z) = \phi$$

where ϕ denotes null set, then the algorithm terminates.

Definition of an Algorithm

Consider a particular nonlinear programming (NLP) problem (10.1). For all $k \geq 1$ define a set $V_k \subset V$ and the algorithmic map

$$A_k\colon \quad V_k \longrightarrow V_{k+1}$$

Then the **algorithm** operates as follows: Given $z^1 \in V_1$, assume $z^2, \ldots, z^k$ have been generated. If

$$A_k(z^k) = \phi$$

the procedure stops. Otherwise, select the successor point z^{k+1} by

$$z^{k+1} \in A_k(z^k)$$

A Sequence as an Algorithm

One value of defining an algorithm in this manner is that any sequence $\{y^k\}_1^\infty$ in V can be reinterpreted as an algorithm. Define the sets

$$V_k = [y^k] \quad \text{for all} \quad k$$

where $[y^k]$ indicates the set consisting only of the point y^k. Let $z^1 = y^1$ and

$$A_k(z^k) = [y^{k+1}] \quad \text{for all} \quad k$$

Then the algorithm so defined generates the sequence $\{y^k\}_1^\infty$.

The preceding paragraph illustrates the generality of the new algorithm definition in that even arbitrary sequences can become algorithms. Moreover, the interpretation of sequences as algorithms will be quite useful in the remaining chapters. As an illustration, suppose a complicated algorithm for the NLP problem generates a sequence of points $\{z^k\}_1^\infty$. To prove convergence of this algorithm, it is often easier to define the sequence $\{z^k\}_1^\infty$ as a second algorithm and establish convergence for the second algorithm. Clearly, convergence of the second algorithm to a solution point implies convergence of the original algorithm.

11.2.1. A Convergent Algorithm

Up to this point we have used the term convergence somewhat imprecisely; now we must define it.

Definition of a Convergent Algorithm

Given a problem and a solution set $\Omega \subset V$ a **convergent algorithm** is an algorithm with the following properties:

(a) If the algorithm stops at a point z, then it indicates either that no solution exists or that z is a solution. Also if a point z is a solution, then either $A_k(z) = \phi$ or $y \in A_k(z)$ implies y is a solution.

(b) Assume the algorithm generates an infinite sequence of points none of which is a solution. Then if all points are not on a compact set, no solution point exists, while if all points are on a compact set, the limit of any convergent subsequence is a solution.

Comments

Part (a) of the convergence definition specifies the behavior of the algorithm should it determine on some iteration k that z^k is a solution. In essence, part (a) requires that the algorithm be able to identify solutions. Simply stated, the algorithm cannot be stupid; it must know a solution when it sees one. This requirement will also be imposed for the sufficient conditions of the next convergence theorem.

Part (b) deals with the case for which no z^k is a solution. Note the use of the compactness; a solution must exist only if all points are on a compact set. This definition weakens the previous requirements that all points be on a compact set, yet maintains sufficient compactness to ensure appropriate behavior of the sequences.

Finally, it might be noted that any definition of convergence is, of necessity, somewhat arbitrary. However, as will be seen, the above definition is reasonable, quite general, and extremely useful.

11.3. CONVERGENCE THEOREMS

The next convergence theorem concerns algorithms that converge in the sense that they satisfy the above convergence definition; however, several concepts must be developed before we can prove it. First we define the function

$$h(z) = \inf \; \{\|z - y\| \, | \, y \in \Omega\}$$

Intuitively, $h(z)$ specifies the distance from z to the closest point in the solution set Ω. The next lemma develops two important properties of h.

Lemma 11.1: *Let* $h(z) = \inf \; \{\|z - y\| \, | \, y \in \Omega\}$. *Then*
(a) $h(z)$ *is continuous, and*
(b) *if* Ω *is a closed set,* $h(z) = 0$ *if and only if* $z \in \Omega$.

PROOF: First (a) will be proved.
Let

$$z^k \longrightarrow z^\infty \tag{11.1}$$

For any $y' \in \Omega$

$$\|z^k - y'\| \leq \|z^\infty - y'\| + \|z^\infty - z^k\|$$

Certainly as y' is arbitrary

$$h(z^k) = \inf \; \{\|z^k - y\| \, | \, y \in \Omega\} \leq \|z^\infty - y'\| + \|z^\infty - z^k\| \tag{11.2}$$

But again because (11.2) holds for any $y' \in \Omega$

$$h(z^k) \leq \inf \; \{\|z^\infty - y\| \, | \, y \in \Omega\} + \|z^\infty - z^k\|$$

Therefore,

$$h(z^k) \leq h(z^\infty) + \|z^\infty - z^k\| \tag{11.3}$$

Similarly,

$$h(z^\infty) \leq h(z^k) + \|z^\infty - z^k\| \tag{11.4}$$

and by (11.3) and (11.4)

$$|h(z^k) - h(z^\infty)| \leq \|z^\infty - z^k\| \tag{11.5}$$

Exploiting (11.1), we find that given $\epsilon > 0$, there exists a P such that $k \geq P$ implies

$$\| z^\infty - z^k \| < \epsilon$$

Therefore for $k \geq P$

$$| h(z^k) - h(z^\infty) | < \epsilon$$

In other words,

$$\lim_{k \to \infty} h(z^k) = h(z^\infty) \tag{11.6}$$

which verifies (a).

Now consider (b). Certainly if $z \in \Omega$, $h(z) = 0$. Suppose $h(z) = 0$. By the definition of inf there is a sequence $\{y^k\}_1^\infty$, $y^k \in \Omega$, that converges to z. Then, employing the closedness of Ω, because $y^k \in \Omega$ for all k, $z \in \Omega$. ◆

The h function plays an important role in the proof of necessity. In one aspect of this proof a sequence of points $\{z^k\}_1^\infty$ will be considered. The sequence will have the property that the limit of any convergent subsequence is a solution. As the next lemma demonstrates, under these conditions

$$\lim_{k \to \infty} h(z^k) = 0$$

Roughly, as k increases z^k will be getting closer and closer to a solution.

Lemma 10.2: *Let a sequence $\{z^k\}_1^\infty$ be contained in a compact set. Suppose for any $\mathcal{K}$ if*

$$z^k \longrightarrow z^\infty \qquad k \in \mathcal{K}$$

that

$$z^\infty \in \Omega$$

Define

$$h(z) = \inf \ \{\| z - y \| | y \in \Omega\}$$

Then,

$$\lim_{k \to \infty} h(z^k) = 0$$

PROOF.   Given any subsequence $\{z^k\}_{\mathcal{K}}$, via compactness there is a $\mathcal{K}^1 \subset \mathcal{K}$ such that

$$z^k \longrightarrow z^\infty \qquad k \in \mathcal{K}^1$$

where because the hypothesis requires $z^\infty \in \Omega$,

$$h(z^\infty) = 0$$

Since the continuity of h was established by Lemma 11.1,

$$\lim_{k \in \mathcal{K}^1} h(z^k) = h(z^\infty) = 0$$

Consequently, because any subsequence of $\{h(z^k)\}$ possesses a subsequence that converges to zero, the entire sequence must converge to zero. ◆

Adaption of Z Function

We must now study how the Z function will be used in the next convergence theorem. As mentioned previously, the requirement in Convergence Theorem A of strict adaption of the Z function will not be needed. Indeed, we even weaken the condition of Convergence Theorem B that

$$Z(z^{k+1}) \geq Z(z^k)$$

The requirement is essentially this: Suppose the algorithm generates z^k. The algorithm may then generate points quite arbitrarily for a finite number of iterations. However, after a finite number of iterations L^k we require that

$$Z(z^l) \geq Z(z^k)$$

for all $l \geq L_k + k$. Here the number L_k may depend upon k and z^k. In words, on any iteration subsequent to L_k iterations after iteration k, the z function must be at least as large as it was on iteration k. Thus for L_k iterations after iteration k the z function is allowed to decrease, but after these L_k iterations no decrease is permitted.

As an example suppose $Z(z^k) = Z^k$ and

$$\{Z^k\} = \{1, 2, 3, 2, 3, 4, 3, 4, 5, 4, 5, 6, 5, 6, 7, 6, 7, 8 \ldots\}$$

Let $L_k = 2$ for all k; then the adaption requirement holds. Observe that $Z^3 = 3$ but $Z^4 = 2$. Thus for $k = 3$

$$Z^{k+1} < Z^k$$

However, for all $k \geq 5$

$$Z^k \geq Z^3$$

Intuitively, the adaption requirement stipulates that the general trend of the sequence is to increase. This requirement is weaker than the strictly monotonic increasing sequence required by Convergence Theorem A.

The following lemma also helps to clarify the adaption requirement. It generalizes Lemma 4.1 and is proved similarly.

Lemma 11.3: *Suppose Z is a continuous function $Z: V \longrightarrow E^1$.*
Assume for any z^k that there is a positive integer L_k such that if $l \geq L_k + k$

$$Z(z^l) \geq Z(z^k) \tag{11.7}$$

Also suppose for some subsequence $\{z^k\}_{\mathscr{K}}$

$$z^k \longrightarrow z^\infty \qquad k \in \mathscr{K} \tag{11.8}$$

Then under these hypotheses

$$\lim_{k \to \infty} Z(z^k) = Z(z^\infty) \tag{11.9}$$

PROOF: For simplicity in the proof we define

$$Z^k = Z(z^k) \quad \text{and} \quad Z^\infty = Z(z^\infty) \tag{11.10}$$

By continuity of Z

$$\lim_{k \in \mathscr{K}} Z^k = Z^\infty \tag{11.11}$$

Then given $\epsilon > 0$, there is an integer P such that if $k' > P$ with $k' \in \mathscr{K}$, and $k'' > P$ with $k'' \in \mathscr{K}$,

$$|Z^{k'} - Z^{k''}| < \epsilon \tag{11.12}$$

Select a particular $k = k'$ such that $k' > P$ with $k' \in \mathscr{K}$, and define $P^1 = L_{k'} + k'$. Via (11.7) for any $l \geq P^1$

$$Z^{k'} \leq Z^l \tag{11.13}$$

Now given any such $l \geq P^1$ for $k'' > L_l + l$ and $k'' \in \mathscr{K}$, again from (11.7)

$$Z^l \leq Z^{k''} \tag{11.14}$$

Consequently from (11.12)

$$|Z^l - Z^{k'}| \leq |Z^{k'} - Z^{k''}| < \epsilon \tag{11.15}$$

But if we take the limit of (11.15) for $k' \in \mathscr{K}$, (11.11) ensures that

$$|Z^l - Z^\infty| < \epsilon \tag{11.16}$$

Summarizing (11.16), given any $\epsilon > 0$ for $l \geq P^1$,

$$|Z^l - Z^\infty| < \epsilon$$

However this is simply the definition of convergence. ◆

Closedness of Algorithmic Map Not Required

The above discussion indicates how Convergence Theorem C will generalize condition 2(a) of Convergence Theorem A, the adaption requirement. It only remains to analyze how to eliminate the closed-map requirement, condition 3 of Convergence Theorem A.

The new condition imposed by Convergence Theorem C is basically as follows: Suppose

$$z^k \longrightarrow z' \quad k \in \mathscr{K}$$

where z' is not a solution. Then there must exist another subsequence

$$z^k \longrightarrow z'' \quad k \in \mathscr{K}'$$

such that

$$Z(z'') > Z(z') \tag{11.17}$$

How does the above condition eliminate the closed-map requirement? Assume the hypotheses of Convergence Theorem A hold. Again suppose

$$z^k \longrightarrow z' \quad k \in \mathscr{K}$$

where z' is not a solution. Then by Convergence Theorem A, there must exist another subsequence, namely $\{z^{k+1}\}_{\mathscr{K}'}$ where $z^{k+1} \in A(z^k)$ and $\mathscr{K}' \subset \mathscr{K}$, such that

$$z^{k+1} \longrightarrow z'' \quad k \in \mathscr{K}'$$

Since z' is not a solution, condition 3 of Convergence Theorem A ensures that the map A is closed at z' and hence that

$$z'' \in A(z') \tag{11.18}$$

But then condition 2(a) implies

$$Z(z'') > Z(z') \tag{11.19}$$

Clearly, (11.17) and (11.19) are identical. Consequently, Convergence Theorem A arrives at the same result as Convergence Theorem C. In brief, Convergence Theorem C retains the essence of the closed-map requirement, specifically relation (11.19), without requiring the closed map.

General Convergence Theorem C. *For a given NLP problem let there be an algorithm that generates points z^k and a solution set $\Omega \subset V$. Consider the following conditions, 1 and 2.*

Condition 1: (a) *If for some z and k, $A_k(z) = \phi$, then the algorithm indicates either that z is a solution or that no solution exists. Should z be a solution, then either $A_k(z) = \phi$ or $y \in A_k(z)$ implies y is a solution.*
(b) *Suppose the algorithm generates an infinite sequence of points none of which is a solution. Then if a solution exists, there is a compact set $X \subset V$ such that $z^k \in X$ for all k.*

Condition 2: *Suppose that the algorithm generates an infinite sequence of points none of which is a solution, and that the points $z^k \in X$, a compact set, for all k. Then there exists a continuous function $Z: X \rightarrow E^1$ such that:*
(a) *Given any point z^k there exists an L_k such that for all $l \geq L_k + k$*

$$Z(z^l) \geq Z(z^k)$$

(b) *Given any convergent subsequence*

$$z^k \longrightarrow z' \qquad k \in \mathscr{K}$$

if z' is not a solution, then either
(i) *there is a k' such that*

$$Z(z^{k'}) > Z(z')$$

or
(ii) *there is a convergent subsequence $\{Z(z^k)\}_{\mathscr{K}'}$ such that†*

$$\lim_{k \in \mathscr{K}'} Z(z^k) > Z(z') \tag{11.20}$$

Then any algorithm which satisfies conditions 1 and 2 satisfies the convergence definition. Moreover, if Ω is closed, any algorithm that converges in accordance with the convergence definition satisfies conditions 1 and 2.

PROOF:

Sufficiency. Assume the algorithm satisfies conditions 1 and 2 of the hypothesis. We must prove that it converges via the convergence definition.

---

†The subset $\mathscr{K}'$ need not be contained in $\mathscr{K}$.

Clearly, condition l(a) and part (a) of the convergence definition are identical.

We must now establish that part (b) of the convergence definition holds. Suppose an infinite sequence is generated no point of which is a solution. If all points are not in a compact set, then by condition 1(b) no solution point exists.

If all points are in a compact set, then it only remains to prove that the limit of any convergent subsequence must be a solution. Suppose

$$z^k \longrightarrow z' \quad k \in \mathscr{K}$$

We will assume z' is not a solution and prove a contradiction. The proof will be given only for condition 2(b) part (ii) as the proof for part (i) is similar.

From Lemma 11.3 and 2(a)

$$\lim_{k \to \infty} Z(z^k) = Z(z') \tag{11.21}$$

Using condition 2(b) part (ii), we find there is a $\mathscr{K}'$ such that (11.20) holds. Then employing compactness, we may find a $\mathscr{K}'' \subset \mathscr{K}'$ such that

$$z^k \longrightarrow z'' \quad k \longrightarrow \mathscr{K}''$$

where since Z is continuous,

$$\lim_{k \in \mathscr{K}''} Z(z^k) = Z(z'') \tag{11.22}$$

But Lemma 11.3 ensures that

$$Z(z'') = \lim_{k \to \infty} Z(z^k)$$

Consequently, by Equation (11.21)

$$Z(z') = Z(z'') \tag{11.23}$$

However, from condition 2(b) part (ii) and (11.22)

$$Z(z'') > Z(z') \tag{11.24}$$

Since (11.23) and (11.24) are contradictory, z' must be a solution.

Necessity. Recall Ω is by hypothesis closed. Assume the algorithm converges, that is, the algorithm satisfies the convergence definition. We must prove that it also satisfies conditions 1 and 2 of the theorem.

Part (a) of the convergence definition implies condition 1(a).

Suppose an infinite sequence of points none of which is a solution is generated. By part (b) of the convergence definition, if a solution exists, all points are in a compact set $X \subset V$. Condition 1(b) holds.

It remains to prove condition 2 in the hypothesis. All points are in a compact set, but no point is a solution. Define the function

$$Z(z) = - \inf \ \{\|z - y\| | y \in \Omega\} = -h(z) \qquad (11.25)$$

where h is defined in Lemma 11.1. From that lemma it is known that Z is a continuous function $Z: V \longrightarrow E^1$ and that

$$Z(z) = 0 \quad \text{if and only if} \quad z \in \Omega \qquad (11.26)$$

as Ω is closed.

Consider the sequence $\{Z(z^k)\}_1^\infty$. Observe:

(a) $Z = -h$

(b) all z^k are in a compact set

and

(c) $z^k \longrightarrow z^\infty, k \in \mathscr{K}$, implies by the convergence definition that $z^\infty \in \Omega$ and consequently by (11.26) that $Z(z^\infty) = 0$.

From (a), (b), and (c) Lemma 11.2 assures us that

$$\lim_{k \to \infty} Z(z^k) = 0 \qquad (11.27)$$

Now examine condition 2(a). Since, by assumption of condition 2, z^k is not a solution

$$Z(z^k) < 0 \qquad (11.28)$$

But by (11.27) there exists an L_k such that $l \geq L_k + k$ implies

$$Z(z^k) < Z(z^l) \leq 0 \qquad (11.29)$$

Condition 2(a) is verified.

Certainly, condition 2(b) holds vacuously as the limit of any convergent subsequence must be a solution. ◆

Comments

Any algorithm that satisfies conditions 1 and 2 of Convergence Theorem C converges in the sense that it satisfies the convergence definition. Conversely, any algorithm that converges in the sense of the convergence definition also satisfies conditions 1 and 2.

Condition 1(a) takes care of two situations. First if the algorithm stops at z^k, then either z^k must be a solution or no solution exists. Presumably the algorithm should indicate which of the two has occurred. Second, if z^k is a solution, either the algorithm must stop or all successor points must be solutions.

Condition 1(b) considers the case of an infinite sequence of points being generated none of which is a solution. (Were any one a solution condition 1(a) would apply.) Essentially, condition 1(b) requires all points to be in a compact set if a solution exists.

Condition 2 assumes that all points are in a compact set. Consequently, when proving that an algorithm satisfies condition 2 we may suppose that the algorithm is generating points in a compact set. Hopefully, this assumption will simplify the task of showing that an algorithm satisfies this condition.

To establish condition 2 we must find an appropriate Z function for the algorithm that is continuous, $Z: V \rightarrow E^1$, and satisfies conditions 2(a) and 2(b).

Condition 2(a) is the adaption requirement previously discussed.

Note that the algorithm need satisfy condition 2(b) only if it generates an infinite sequence of points none of which is a solution. Also the algorithm need satisfy 2(b) only for z' not a solution. If z' is already a solution, then condition 2(b) holds vacuously. Furthermore, either part (i) or part (ii), not both, need be satisfied by the algorithm. Part (ii) is the condition that replaced the closed-map assumption, as previously mentioned. Part (i) is a similar condition that also obviates the need for a closed map.

One final convergence theorem, phrased to avoid compactness considerations, will be posed.

Convergence Theorem D. *For a given NLP problem let an algorithm generate points z^k and assume a solution set $\Omega \subset V$ is given.*

Suppose:

Condition 1:   If $A_k(z) = \phi$, then either z is a solution or no solution exists.

Condition 2:   There is a continuous function $Z: V \rightarrow E^1$ such that:

(a) Given any point z^k there exists an L_k such that for all $l \geq L_k + k$ if there is a z^l

$$Z(z^l) \geq Z(z^k)$$

(b) If there exists a convergent subsequence

$$z^k \longrightarrow z' \qquad k \in \mathcal{K}$$

such that z' is not a solution then either
(i) there is a k' such that

$$Z(z^{k'}) > Z(z')$$

or

(ii) *there is a convergent subsequence* $\{Z(z^k)\}_{\mathscr{K}'}$, *such that*

$$\lim_{k \in \mathscr{K}'} Z(z^k) > Z(z')$$

Under these conditions, if the algorithm terminates at z^k, *either* z^k *is a solution or no solution exists. Moreover, should an infinite number of points be generated, if there is a convergent subsequence*

$$z^k \longrightarrow z^\infty \qquad k \in \mathscr{K}$$

then z^∞ *is a solution.*

(Observe for part (ii) that $\mathscr{K}'$ need not be contained in $\mathscr{K}$.)

PROOF: The proof is left as exercise 11.1.

Comment

Note that an algorithm which satisfies the hypotheses of Convergence Theorem D need not fit the convergence definition. The difficulty is essentially one of compactness. Regrettably, it is not an easy task to reformulate the definition of convergence to make the conditions of Convergence Theorem D necessary for convergence. For this reason we have not altered the convergence definition to include algorithms that converge under Convergence Theorem D. Moreover to do so would invalidate the powerful necessary and sufficient conditions of Convergence Theorem C. Nevertheless, an algorithm that satisfies Convergence Theorem D is considered convergent.

Other convergence theorems appear in the exercises. In particular see exercise 11.3 for finite convergence.

EXERCISES

11.1. Prove Convergence Theorem D.

11.2. Prove that Convergence Theorems A and B are special cases of Convergence Theorem C.

11.3. *Finite Convergence of Procedures such as the Simplex Method.* Let V be a finite set; that is, V consists only of a finite number of points.

(a) Let $z^k \longrightarrow z^\infty$, $k \in \mathcal{K}$. Prove that there must exist a P such that for all $k \geq P$ and $k \in \mathcal{K}$

$$z^k = z^\infty$$

(b) Prove the set V is compact.

(c) Prove if an algorithm satisfies the hypotheses of either Convergence Theorem A, B, C or D that the algorithm determines a solution point in a finite number of steps.

Hint: Compactness of V ensures that a convergent subsequence exists. Therefore the limit z^∞ of a convergent subsequence is a solution. Use (a) to prove that z^∞ must be reached after a finite number of iterations.

(d) Prove convergence of the linear-simplex method by letting V be the vertices of the polyhedral set.

(e) Let $Z:$ $V \longrightarrow E^1$ be a function. Prove Z is continuous.

(f) Prove the following theorem, and also show why it is a special case of Convergence Theorems C and D.

Theorem: *Consider an algorithm on a finite set V, with the following properties.*

1: *If z^k is a solution the algorithm terminates.*

2: *There is a function $Z:$ $V \longrightarrow E^1$ such that if z^k is not a solution then there is an L_k for which all $l \geq L_k + k$ imply*

$$Z(z^l) > Z(z^k)$$

Then the algorithm terminates in a finite number of steps.

11.4. Prove the following theorem.

Theorem: *For a set $V \subset E^n$ suppose there is an algorithm with map $A:$ $V \longrightarrow V$ that given $z^1 \in V$ generates a sequence $\{z^k\}_1^\infty$ via the recursion $z^{k+1} \in A(z^k)$. A solution set $\Omega \subset V$ is also given.*

Suppose there exists a continous function $Z:$ $V \longrightarrow E^1$ such that

1: *If $y \in A(z)$ $Z(y) \geq Z(z)$*

2: *If z' is not a solution, there exists a $\delta > 0$ and a $\epsilon > 0$ such that, if $\| z' - z \| < \delta$ then $y \in A(z)$ implies*

$$Z(y) - Z(z) > \epsilon$$

Then under these hypotheses if $z^k \longrightarrow z^\infty$, $k \in \mathcal{K}$, z^∞ is a solution.

11.5. Prove the following theorem.

Theorem: *For $V \subset E^n$ let an algorithm with map $A:$ $V \longrightarrow V$ generate the sequence $\{z^k\}_1^\infty$. Suppose a solution set $\Omega \subset V$ is also specified. Assume*

1: $\lim\limits_{k\to\infty} z^k = z^\infty$

2: *If z' is not a solution, there is a $\delta > 0$ and a $\epsilon > 0$ such that if $\| z' - z \|$ $< \delta$ then $y \in A(z)$ implies*

$$\| y - z \| > \epsilon$$

Then under these hypotheses z^∞ is a solution.
Hint: Let $Z(z) = \| z - z^\infty \|$.

11.6. Prove in Convergence Theorems C and D that condition 2(a) may be replaced by

$$Z(z^{k+1}) \geq Z(z^k) \quad \text{for all} \quad k$$

NOTES AND REFERENCES

This chapter is an exposition of the concepts presented in the paper by ZANGWILL (1966g).

Exercise 11.4 was adapted from a paper by POLAK and DEPARIS that applies convergence theory to an interesting optimal-control algorithm.

12 Penalty and Barrier Methods

Penalty and barrier methods solve the nonlinear programming (NLP) problem by solving a sequence of unconstrained problems, that in effect, suppress the constraints. Because penalty and barrier methods do not consider the constraints explicitly, they have been found to be computationally efficient for the NLP problem with nonlinear constraints. In this chapter we demonstrate how the penalty and barrier methods arise out of a certain unconstrained problem, which is equivalent to the NLP problem, yet cannot be directly solved. Instead the penalty and barrier methods approximate this problem with a sequence of related unconstrained problems.

In addition to presenting the two methods, we analyze their duality properties. The penalty method is particularly interesting in this regard as it generates a dual problem fundamentally different from that considered in Chap. 2. Essentially, a new generalized form of Lagrangean is introduced.

12.1. THE BASIS OF PENALTY AND BARRIER METHODS

Penalty and barrier methods seek a solution to the NLP problem

$$\begin{aligned} \max \quad & f(x) \\ \text{subject to} \quad & g_i(x) \geq 0 \qquad i = 1, \ldots, m \end{aligned} \qquad (12.1)$$

They transform problem (12.1), a problem with constraints, into a sequence of problems each without constraints. Were problem (12.1) to have no constraints, i.e., if $m = 0$, then our goal would be to solve the unconstrained problem

$$\max_{x \in E^n} \ f(x)$$

To solve problem (12.1) for $m > 0$, the penalty and barrier methods introduce the effect of the constraints by modifying the objective function f.

As an illustration of this approach consider the function

$$P^1(x) = \begin{cases} 0 & x \in F \\ -\infty & x \notin F \end{cases} \tag{12.2}$$

where F is the feasible set.

Combine f and P^1 to form

$$C(x) = f(x) + P^1(x) \tag{12.3}$$

Intuitively $P^1(x)$ produces an infinite penalty for leaving the feasible region (see Fig. 12.1). Clearly x^* is optimal for problem (12.1) if and only if x^* also solves the unconstrained problem

$$\max_{x \in E^n} C(x) \qquad\qquad (12.4)$$

Problem (12.1) has thus been transformed into an equivalent unconstrained problem.

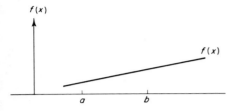

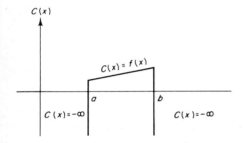

The problem is to maximize f on $[a, b]$.

The feasible region is the line segment $[a, b]$.

Fig. 12.1.
An infinite penalty.

Regrettably, this particular unconstrained problem is unsolvable in a practical sense. The value $-\infty$ is impossible to calculate on a computer. Moreover, suppose in place of (12.2) we were to define

$$P^2(x) = \begin{cases} 0 & x \in F \\ -L & x \notin F \end{cases} \qquad\qquad (12.5)$$

where L is very large, and to define $C(x)$ as in (12.3) but with P^2 replacing P^1. The resulting $C(x)$ function is still extremely difficult to maximize due to its severe discontinuities, very large L, and the resultant round-off error in the computer.

Penalty and barrier methods approximate P^1 by a sequence of functions that, in the limit, approach P^1. By choosing these functions astutely, say continuous and differentiable, the methods circumvent the computational difficulties with P^1 or P^2.

Penalty Method Motivation

To develop the penalty method consider first a special case of the penalty function

$$P^3(x) = -\sum_{i=1}^{m} \text{min} \quad [g_i(x), 0]^2 \qquad (12.6)$$

Certainly

$$P^3(x) = 0 \quad \text{if and only if} \quad x \in F \qquad (12.7)$$

Moreover, if $r > 0$, the function

$$\frac{1}{r}P^3(x)$$

approximates the $P^1(x)$ function as $r \longrightarrow 0$ (see Fig. 12.2).

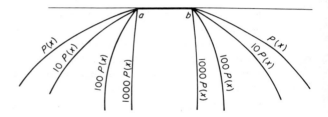

Fig. 12.2.
Representation of $\frac{1}{r}P(x)$.

The penalty method uses a sequence $\{r^k\}_1^\infty$ such that $r^k > 0$ for all k, $r^k > r^{k+1}$, and

$$\lim_{k \to \infty} \quad r^k = 0$$

Define x^k as an x that maximizes

$$C(x, r^k) = f(x) + \frac{1}{r^k}P^3(x) \qquad (12.8)$$

(We assume these x^k exist.) Specifically,

$$C(x^k, r^k) = \max_{x \in E^n} C(x, r^k)$$

Then we determine an optimal point via the penalty method as follows: Given $r^1 > 0$, where r^1 is usually chosen to be 1, solve for x^1 using

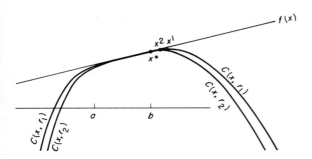

Fig. 12.3.

Penalty methods. The points x^1 and x^2 are indicated as well as the optimal point x^*. Clearly, $\lim x^k = x^*$.

an unconstrained maximization technique. Then, using r^2, we solve for x^2, again using an unconstrained maximization technique. In this manner the sequence $\{x^k\}_1^\infty$ is generated. As will be proved, a limit of any convergent subsequence of $\{x^k\}_1^\infty$ is an optimal point for problem (12.1) (see Fig. 12.3).

The precise proof uses a more general form for $C(x, r)$ than that in (12.8). Also note that the successive calculation of x^{k+1} from x^k may not be difficult. For suppose x^k has just been calculated. If r^{k+1} is not too much smaller than r^k, the point x^{k+1} should be near x^k. By starting the unconstrained technique from x^k, we may easily obtain the new point x^{k+1}.

In general, penalty methods approximate the $P^1(x)$ of (12.2) by a function $(1/r)P(x)$ that approaches $P^1(x)$ from outside the feasible region as $r \rightarrow 0$. Essentially $P(x)$ imposes a penalty for leaving the feasible region.

Barrier Method Motivation

Barrier methods approximate $P^1(x)$ by functions $rB(x)$ that approach $P^1(x)$ from inside the feasible region. Roughly, $B(x)$ sets up a barrier against leaving the feasible region. For example, let

$$B(x) = -\sum_{i=1}^m \frac{1}{g_i(x)} \quad \text{where} \quad g_i(x) > 0 \qquad i = 1, \ldots, m \qquad (12.9)$$

Clearly $B(x) \rightarrow -\infty$ as $g(x) \rightarrow 0$. Thus as x gets close to the boundary of the feasible region, $B(x)$ becomes infinitely small.

Consider the function

$$D(x, r) = f(x) + rB(x) \qquad (12.10)$$

and a sequence $\{r^k\}_1^\infty$ such that

$$r^k > r^{k+1} \quad \text{and} \quad \lim_{k \to \infty} r^k = 0 \tag{12.11}$$

(Note the use of r instead of $1/r$ as in the penalty method.) The barrier method determines a sequence $\{x^k\}_1^\infty$ such that x^k maximizes $D(x, r^k)$ over F.† We will later prove that any limit point of this sequence is optimal for problem (12.1) (see Fig. 12.4).

Although the problem of maximizing $D(x, r^k)$ over F is a constrained problem, by attacking it in a special manner, we can actually solve it using an unconstrained search technique. To determine x^1 start the unconstrained search from an initial feasible point x^0 such that

$$g_i(x^0) > 0 \quad i = 1, \ldots, m$$

Then, because as the boundary of F is approached, $B(x) \to \infty$, the points generated by an unconstrained optimization technique will remain inside the feasible region and determine x^1. Moreover, x^1 will satisfy

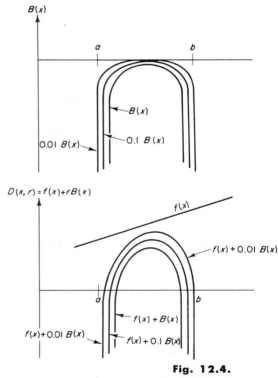

Fig. 12.4.

Barrier Method.

---

†In Sec. 12.3 the maximum is defined only over the interior of F. However, as will be shown in that section, the maxima are the same.

$$g_i(x^1) > 0 \qquad i = 1, \ldots, m$$

Similarly, the point x^2 is obtained using an unconstrained search that starts from x^1. If we continue in this manner, because $B(x)$ imposes a barrier to leaving the feasible region, an unconstrained search will determine all points x^k. Furthermore, the x^k generated will all be on the interior of F, that is, for all k

$$g_i(x^k) > 0 \qquad i = 1, \ldots, m$$

With the motivation of the penalty and barrier methods completed, we turn to a detailed discussion and proof of the penalty method.

12.2. PENALTY METHODS

Let $\{r^k\}_1^\infty$ be a sequence

$$r^k > 0 \qquad r^k > r^{k+1}$$

such that

$$\lim_{k \to \infty} r^k = 0 \tag{12.12}$$

Define a continuous function $P(x)$, termed a penalty function, such that

$$P(x) \begin{cases} = 0 & \text{if} \quad x \in F \\ < 0 & \text{if} \quad x \notin F \end{cases} \tag{12.13}$$

A typical form of P is

$$P(x) = -\sum_{i=1}^m |\min \ [g_i(x), 0]|^{1+\epsilon} \tag{12.14}$$

where $\epsilon \geq 0$, although many other forms are possible.
Define

$$C(x, r) = f(x) + \frac{1}{r} P(x) \tag{12.15}$$

for $r > 0$.

The Algorithm

The penalty procedure is as follows: For r^k calculate x^k such that

$$C(x^k, r^k) = \max_{x \in E^n} C(x, r^k) \tag{12.16}$$

Should x^k be optimal for problem (12.1), stop; otherwise, continue with r^{k+1}.

Convergence

To prove convergence of the algorithm we assume that the f and g_i are continuous and that an optimal point exists. Also suppose that the maxima are achieved so that all x^k exist, and that there is a compact set X such that $x^k \in X$ for all k. The function $P(x)$ can usually be selected to ensure that all x^k and the set X exist, especially when F is compact.

Convergence will follow easily from the next lemma.

Lemma 12.1:

$$C(x^k, r^k) \geq C(x^{k+1}, r^{k+1}) \tag{12.17}$$

$$P(x^k) \leq P(x^{k+1}) \tag{12.18}$$

$$f(x^k) \geq f(x^{k+1}) \tag{12.19}$$

PROOF.   From (12.13)

$$P(x) \leq 0 \quad \text{for all} \quad x$$

Because $1/r^k < 1/r^{k+1}$,

$$f(x^{k+1}) + \frac{1}{r^k}P(x^{k+1}) \geq f(x^{k+1}) + \frac{1}{r^{k+1}}P(x^{k+1})$$

Furthermore, since x^k maximizes $f(x) + r^k P(x)$,

$$C(x^k, r^k) = f(x^k) + \frac{1}{r^k}P(x^k) \geq f(x^{k+1}) + \frac{1}{r^k}P(x^{k+1}) \geq C(x^{k+1}, r^{k+1})$$

which proves (12.17).

Also because x^{k+1} maximizes $C(x, r^{k+1})$,

$$f(x^k) + \frac{1}{r^{k+1}}P(x^k) \leq f(x^{k+1}) + \frac{1}{r^{k+1}}P(x^{k+1})$$

Similarly

$$f(x^{k+1}) + \frac{1}{r^k}P(x^{k+1}) \leq f(x^k) + \frac{1}{r^k}P(x^k) \tag{12.20}$$

Adding and rearranging

$$\frac{1}{r^{k+1}}[P(x^k) - P(x^{k+1})] \le \frac{1}{r^k}[P(x^k) - P(x^{k+1})]$$

But as $1/r^{k+1} > 1/r^k$

$$P(x^k) \le P(x^{k+1}) \qquad\qquad (12.21)$$

Finally, from (12.20) and (12.21)

$$f(x^{k+1}) - f(x^k) \le \frac{1}{r^k}[P(x^k) - P(x^{k+1})] \le 0 \quad \blacklozenge$$

Convergence will be established using Convergence Theorem C.

Let $z = x$. Recall from Chap. 11 that any sequence may be defined as an algorithm. Suppose the penalty method generates the sequence $\{x^k\}_1^\infty$. We now define the sequence $\{x^k\}_1^\infty$ as an algorithm, and prove its convergence. Thus we will establish that any limit point of any convergent subsequence of $\{x^k\}_1^\infty$ is a solution.

Condition 1

A point x^* is termed a solution if x^* is optimal for problem (12.1). Clearly, condition 1(a) of Convergence Theorem C holds. By assumption all x^k are on a compact set X, and a solution exists so that condition 1(b) holds.

Condition 2

Define $Z(x) = -f(x)$. Via Lemma 12.1

$$Z(x^{k+1}) = -f(x^{k+1}) \ge -f(x^k) = Z(x^k) \qquad\qquad (12.22)$$

Therefore condition 2(a) holds by letting $L_k = 1$ for all k.

To prove condition 2(b), suppose $x^k \longrightarrow x'$, $k \in \mathcal{K}$. Observe, because the optimal point $x^* \in F$, by (12.13) and definition of x^k

$$f(x^*) = C(x^*, r^k) \le C(x^k, r^k) = f(x^k) + \frac{1}{r^k}P(x^k) \qquad (12.23)$$

From Lemma 12.1 and (12.23) the sequence $\{C(x^k, r^k)\}_1^\infty$ is nonincreasing and is bounded below by $f(x^*)$. It therefore has a limit (exercise 12.1)

$$C^\infty = \lim_{k \to \infty} C(x^k, r^k) = \lim_{k \in \mathcal{K}} C(x^k, r^k) \qquad (12.24)$$

Exploiting the continuity of f, we find that

$$\lim_{k \in \mathscr{K}} f(x^k) = f(x') \tag{12.25}$$

and using (12.24) and (12.25), we arrive at

$$\lim_{k \in \mathscr{K}} \frac{1}{r^k} P(x^k) = \lim_{k \in \mathscr{K}} C(x^k, r^k) - \lim_{k \in \mathscr{K}} f(x^k) = C^{\infty} - f(x') \tag{12.26}$$

But $P(x^k) \leq 0$ and $\lim 1/r^k \to +\infty$, so that (12.26) would hold only if

$$\lim_{k \in \mathscr{K}} P(x^k) = 0 \tag{12.27}$$

Via the continuity of $P(x^k)$ and (12.27)

$$\lim_{k \in \mathscr{K}} P(x^k) = P(x') = 0 \tag{12.28}$$

Hence, we arrive at the important conclusion that $x' \in F$.

One more result is required. Again using (12.23), since $1/r^k \, P(x^k) \leq 0$, we find

$$f(x^k) \geq f(x^*) \tag{12.29}$$

Then continuity of f and (12.29) imply

$$\lim_{k \in \mathscr{K}} f(x^k) = f(x') \geq f(x^*)$$

We have thus proved that $x' \in F$ and $f(x') \geq f(x^*)$. Consequently, the point x' must be a solution. The condition 2(b) holds vacuously as x' must be a solution, and the hypotheses of Convergence Theorem C have been established.

It should be noted that Convergence Theorem C provided little more than a hollow shell for this convergence proof. Proving that x' was optimal actually established the proof.

Also observe that (exercise 12.2)

$$\lim_{k \to \infty} C(x^k, r^k) = f(x^*) \tag{12.30}$$

An Example

Consider the following illustrative problem:

$$\text{max} \quad -x^2 + 10x$$
$$\text{subject to} \quad x \leq 1$$

The optimal point is clearly $x^* = 1$.

In order to solve this problem by the penalty algorithm, use the quadratic-loss function

$$P(x) = -(\min \quad [0, 1 - x])^2$$

Then $C(x, r) = -x^2 + 10x - 1/r(\quad \min \quad [0, 1 - x])^2$. Since $C(x, r)$ is concave and continuously differentiable, its maximum can be found by differentiation. The corresponding x that maximizes $C(x, r)$ as $r \to 0$ will be the optimal point.

$$\frac{\partial C(x, r)}{\partial x} = -2x + 10 + 2\frac{1}{r}(\min \quad [0, 1 - x]) = 0$$

or

$$\min \quad \left[-2x + 10, \; -2x\left(1 + \frac{1}{r}\right) + 10 + 2\frac{1}{r}\right] = 0$$

If

$$-2x + 10 = 0$$

then

$$x = 5$$

If

$$-2x\left(1 + \frac{1}{r}\right) + 10 + 2\frac{1}{r} = 0$$

then

$$x = \frac{10r + 2}{2(1 + r)}$$

By evaluating the minimum at the values $x = 5$ and $x = (10r + 2)/2(1 + r)$, we can easily check that for all $r > 0$ the minimum achieved for $x = (10r + 2)/2(1 + r)$. The algorithm requires $r \to 0$, and the optimal point is thus $x^* = 1$. Note that the point $x = 5$ yields the unconstrained maximum of the objective function $-x^2 + 10x$.

12.2.1. Duality

Associated with the general NLP problem are various dual problems. This section develops a new form of the dual problem, somewhat different from that developed in Chap. 2, and shows the points (x^k, r^k) in E^{n+1}, where x^k maximizes $C(x, r^k)$, are feasible for this new dual problem. Thus the algorithm is a dual method.

The following two problems can be posed:

Primal problem:
$$\sup_{x \in E^n} \inf_{r \in R} \; C(x, r)$$

Dual problem:
$$\inf_{r \in R} \sup_{x \in E^n} \; C(x, r)$$

where $R = (0, +\infty)$.

Primal Problem

Since $P(x)$ is zero if x is feasible and is negative otherwise, the primal problem can be written

$$\sup_{x \in E^n} \inf_{r \in R} \; C(x, r) = \sup \; \{f(x) \quad \text{where} \quad P(x) = 0\}$$
$$= \begin{cases} \sup \; f(x) \\ \text{subject to} \\ g_i(x) \geq 0 \qquad i = 1, \ldots, m \end{cases}$$

Thus the primal problem is the NLP problem. As previously assumed, the maximum operation can replace the supremum.

Dual Problem

We now prove that the dual problem, under the previous assumptions, is solved by the penalty algorithm. It was assumed that

$$\sup_{x \in E^n} \; C(x, r) = \max_{x \in E^n} \; C(x, r)$$

By Lemma 12.1

$$\max_{x \in E^n} \; C(x, r^{k+1}) \leq \max_{x \in E^n} \; C(x, r^k)$$

Thus

$$\inf_{r \in R} \sup_{x \in E^n} \; C(x, r) = \inf_{r \in R} \max_{x \in E^n} \; C(x, r) = \lim_{k \to \infty} C(x^k, r^k)$$

But, as shown in (12.30),

$$\lim_{k \to \infty} C(x^k, r^k) = f(x^*)$$

$$= \max \quad f(x)$$

$$\text{subject to} \quad g_i(x) \geq 0 \qquad i = 1, \ldots, m$$

Summarizing the dual problem, we arrive at

$$\inf_{r} \sup_{x} \quad C(x, r) = \max \quad f(x)$$

$$\text{subject to} \quad g_i(x) \geq 0 \qquad i = 1, \ldots, m$$

Saddle Value

Consequently, the value of the primal problem is the value of the dual problem and

$$\inf_{r \in R} \sup_{x \in E^n} \quad C(x, r) = \sup_{x \in E^n} \inf_{r \in R} \quad C(x, r)$$

It is interesting to note that the development of the primal and dual problems thus far has been for the general NLP problem, and no concavity or convexity has been imposed. Dual equality thus holds for some very general functions. To obtain further results, however, we must impose a concavity assumption.

Concavity

Consider now the NLP problem with f and g_i concave and continuously differentiable. $C(x, r)$ can also be chosen to be concave and continuously differentiable (exercise 12.3); then

$$C(x^1, r^1) = \max_{x} \quad C(x, r^1)$$

if and only if

$$\nabla f(x^1) + \frac{1}{r^1} \nabla P(x^1) = 0$$

The dual problem can thus be rewritten

$$\inf_{x, r} \quad f(x) + \frac{1}{r} P(x)$$

$$\text{subject to} \quad \nabla f(x) + \frac{1}{r} \nabla P(x) = 0$$

$$r \geq 0$$

or, equivalently, the dual problem becomes

$$\lim_{k \to \infty} f(x) + \frac{1}{r^k} P(x)$$

$$\text{subject to} \quad \nabla f(x) + \frac{1}{r^k} \nabla P(x) = 0$$

Depending upon the loss function involved, we can develop a large number of dual problems. For example, consider the function $P(x)$ given by

$$P(x) = -\sum_{i=1}^{m} |\quad \min \quad [g_i(x), 0]|^{1+\epsilon} \quad \text{for} \quad \epsilon \geq 0$$

If $\epsilon > 0$, then $\nabla P(x)$ is continuous;

$$\frac{\partial P(x)}{\partial x_j} = \sum_{i=1}^{m} (1 + \epsilon)(-\min \quad [g_i(x), 0])^{\epsilon} \frac{\partial g_i(x)}{\partial x_j} \quad j = 1, \ldots, n$$

Define

$$u_i = \frac{1}{r}(-\min \quad [g_i(x), 0])^{\epsilon}$$

The dual problem can then be written, for $\epsilon > 0$ and fixed,

$$\inf_{x, r} \quad f(x) + \sum u_i g_i(x)$$

$$\nabla f(x) + (1 + \epsilon) \sum u_i \nabla g_i(x) = 0$$

$$u_i - \frac{1}{r}(-\min \quad [g_i(x), 0])^{\epsilon} = 0$$

$$r \geq 0$$

Note the relationship between this dual problem and the concave dual developed in Chap 2,

$$\inf_{x, \lambda} \quad f(x) + \sum_{i=1}^{m} \lambda_i g_i(x)$$

$$\nabla f(x) + \sum_{i=1}^{m} \lambda_i \nabla g_i(x) = 0$$

$$\lambda_i \geq 0$$

where $\lambda = (\lambda_1, \ldots, \lambda_m)$.

12.3. A BARRIER METHOD

As implied previously, the essence of the barrier method is embodied in the barrier function $B(x)$. However, to study the barrier function $B(x)$, we must first define some concepts. The **interior** of the feasible region is the set

$$F^0 = \{x \mid g_i(x) > 0 \qquad i = 1, \ldots, m\}$$

With F compact and continuous constraints the interior F^0 will not be a closed set. A point x is on the **boundary** of the feasible region if $x \in F$ and

$$g_j(x) = 0 \quad \text{for some} \quad j$$

Clearly F consists of both its interior F^0 and its boundary.

Definition of the Barrier Function

For each $x \in F^0$ define $B(x)$ as a continuous function such that

(a) $B(x) \leq 0$ (12.31)

and

(b) $B(x) \longrightarrow -\infty$ as x approaches the boundary of F (12.32)

Another way of phrasing (b) is that if $y^k \longrightarrow y^\infty$, $k \in \mathcal{K}$, where $y^k \in F^0$ for all k and if for some j

$$g_j(y^\infty) = 0 \tag{12.33}$$

then

$$\lim_{k \in \mathcal{K}} B(y^k) \longrightarrow -\infty$$

An example of B was given in Sec. 12.1.

Note that, since F^0 is not a closed set we may find a subsequence $y^k \in F^0$

$$y^k \longrightarrow y^\infty \qquad k \in \mathcal{K}$$

where y^∞ is in the boundary of F but is not in F^0. Observe also that B is continuous on F^0 but is not even defined on the boundary of F.

Now define the function for $x \in F^0$

$$D(x, r) = f(x) + rB(x) \quad \text{where} \quad r > 0 \tag{12.34}$$

We assume f is continuous, then $D(x, r)$ is continuous on F^0.

The Subproblem

The barrier method requires solution of the subproblem for $r > 0$

$$\sup_{x \in F^0} D(x, r) \tag{12.35}$$

We now show that the maximizing point actually exists; then the maximum operation may be used instead of supremum. The difficulty is that, even though F is compact, F^0 is not, hence, we cannot apply the results of Chap. 8. Instead the definition of supremum must be invoked. Suppose

$$\sup_{x \in F^0} D(x, r) = \gamma \tag{12.36}$$

Then by definition of supremum there is a sequence $y^k \in F^0$ such that

$$\lim_{k \to \infty} D(y^k, r) = \gamma$$

Since the y^k are also in F, by the compactness of F a convergent subsequence $\{y^k\}_{\mathscr{K}}$ exists

$$\lim_{k \in \mathscr{K}} y^k = y^\infty$$

where $y^\infty \in F$.

First assume $y^\infty \in F^0$, then as D is continuous on F^0

$$\gamma = \lim_{k \in \mathscr{K}} D(y^k, r) = \lim_{k \in \mathscr{K}} f(y^k) + r \lim_{k \in \mathscr{K}} B(y^k) = f(y^\infty) + rB(y^\infty)$$

and y^∞ achieves the maximum.

Suppose now that y^∞ is not in F^0 but is on the boundary of F. Then

$$\gamma = \lim_{k \in \mathscr{K}} D(y^k, r) = f(y^\infty) + r \lim_{k \in \mathscr{K}} B(y^k) \tag{12.37}$$

But Equation (12.37) cannot hold, because $r > 0$ and from property (b) of the barrier function $\lim_{k \in \mathscr{K}} B(y^k) \to -\infty$. Therefore we arrive at a contradiction, and y^∞ cannot be on the boundary of F.

Consequently, it must be that $y^\infty \in F^0$ and we may replace the supremum in (12.35) by maximum.

The Barrier Algorithm

The barrier method will now be stated. A sequence $\{r^k\}_1^\infty$ is given such that $r^k > 0$, $r^k > r^{k+1}$, and $\lim\limits_{k \to \infty} r^k = 0$. From a point $x^0 \in F^0$, assumed to exist, calculate

$$D(x^1, r^1) = \max_{x \in F^0} \ D(x, r^1)$$

And in general, starting from $x^k \in F^0$, we calculate x^{k+1} to

$$\max_{x \in F^0} \ D(x, r^{k+1})$$

As mentioned in Sec. 12.1, we can perform the constrained maximization, $\max\limits_{x \in F^0} \ D(x, r^k)$, by an unconstrained optimization technique starting from x^{k-1}. Also $x^k \in F^0$ for all k.

Convergence

The convergence proof requires the following assumptions, some of which have already been stated but are reiterated here for clarity. The functions f and g_i are continuous, the feasible set F is compact, and there exists a point $x^0 \in F^0$. Also assume that if x^* is optimal for problem (12.1) then in any neighborhood of x^* there is a $y \in F^0$. This final assumption requires that x^* cannot be isolated from points in F^0.

To prove convergence we must first state the following lemma whose proof is similar to that of Lemma 12.1 and is left as exercise 12.10.

Lemma 12.2:

$$f(x^k) \leq f(x^{k+1}) \tag{12.38}$$

$$B(x^k) \leq B(x^{k+1}) \tag{12.39}$$

$$D(x^k, r^k) \leq D(x^{k+1}, r^{k+1}) \tag{12.40}$$

Convergence of the barrier method will be established using Convergence Theorem C. We define the algorithm by the sequence $\{x^k\}_1^\infty$, and a point x^* is termed a solution if x^* is optimal for problem (12.1). Condition 1(a) holds as all points x^k are in the feasible set, and by (12.38), if x^k is optimal, all successors are optimal.

Condition 1(b) is verified by noting that all points are in the compact set F while f, being continuous on the compact set F, ensures that an optimal point exists.

For condition 2 let $Z(x) = f(x)$. Via (12.38) condition 2(a) is verified. To prove (2b), suppose

$$x^k \longrightarrow x' \qquad k \in \mathcal{K}$$

where x' is not a solution. Then, as x' must be feasible,

$$f(x') < f(x^*) \tag{12.41}$$

By assumption we may find a point $y \in F^0$ so near to x^* that

$$f(y) > f(x') \tag{12.42}$$

Moreover, because $y \in F^0$, $B(y)$ is defined and finite. Employing the fact that $\lim r^k \rightarrow 0$, we may pick k so large, say $k = k'$, that

$$D(y, r^{k'}) = f(y) + r^{k'}B(y) > f(x') \tag{12.43}$$

Then

$$D(x^{k'}, r^{k'}) = \max_{x \in F^0} D(x, r^{k'}) > f(x') \tag{12.44}$$

Moreover, since

$$r^{k'}B(x^{k'}) < 0$$

(12.44) provides the conclusion that

$$f(x^{k'}) > f(x') - r^{k'}B(x^{k'}) > f(x') \tag{12.45}$$

This verifies condition 2(b), and the procedure satisfies Convergence Theorem C.

Corollary 12.2.1:

$$\lim_{k \to \infty} r^k B(x^k) = 0$$

(exercise 12.5).

12.3.2. Concave Duality for Barrier Methods

Suppose that

$$B(x) = -\sum_{i=1}^{m} \frac{1}{g_i(x)}$$

and that f and g_i are continuously differentiable and concave. Then because $-1/g_i(x)$ is concave for $x \in F^0$ (exercise 12.6).

$$D(x, r) = f(x) - r \sum_{i=1}^{m} \frac{1}{g_i(x)}$$

is concave on F^0. Consequently, as x^k maximizes $D(x, r^k)$

$$0 = \nabla D(x^k, r^k) = \nabla f(x^k) + r^k \sum_{i=1}^{m} \frac{1}{g_i(x^k)^2} \nabla g_i(x^k) \qquad (12.46)$$

Now define

$$\lambda_i^k = \frac{r^k}{g_i(x^k)^2}$$

Then the Lagrangean becomes

$$L(x^k, \lambda^k) = f(x^k) + \sum_{i=1}^{m} \lambda_i^k g_i(x^k) = f(x^k) + r^k \sum_{i=1}^{m} \frac{1}{g_i(x^k)} \qquad (12.47)$$

and

$$\nabla_x L(x^k, \lambda^k) = \nabla f(x^k) + \sum \lambda_i^k \nabla g_i(x^k) = \nabla D(x^k, r^k) \qquad (12.48)$$

where the final equality holds via (12.46).

The Dual Problem

From Chap. 2 the dual programming problem, because the f and g_i are concave, becomes

$$\min \quad L(x, \lambda)$$
$$\text{subject to} \quad \nabla L(x, \lambda) = 0$$
$$\lambda \geq 0$$

But using (12.46) and (12.48) we observe that the points (x^k, λ^k) are dual feasible. Therefore, the barrier algorithm not only generates primal feasible points x^k but also determines dual feasible points (x^k, λ^k).

Error Bounds

Duality theory ensures that the function

$$L(x^k, \lambda^k) = f(x^k) + r^k \sum_{i=1}^{m} \frac{1}{g_i(x^k)}$$

is an upper bound to $f(x^*)$. Furthermore, because x^k is feasible and x^* is optimal,

$$f(x^k) \leq f(x^*) \leq f(x^k) + r^k \sum_{i=1}^{m} \frac{1}{g_i(x^k)}$$

Consequently,

$$r^k \sum_{i=1}^{m} \frac{1}{g_i(x^k)}$$

may be considered as an error term as it is a bound on the difference between $f(x^k)$ and $f(x^*)$.

Via Corollary 12.2.1

$$\lim_{k \to \infty} L(x^k, r^k) \longrightarrow f(x^*)$$

and the error term goes to zero in the limit.

EXERCISES

12.1. Consider a sequence $\{C^k\}_1^\infty$ in E^1. Suppose the sequence is nonincreasing and bounded below by L. That is,

$$C^k \geq C^{k+1} \quad \text{for all} \quad k$$

and

$$C^k \geq L \qquad \text{for all} \quad k$$

Prove the sequence converges: $\lim_{k \to \infty} C^k = C^\infty$.

Hint: All points are in the compact set $[L, C^1]$. A convergent subsequence has a limit. Prove the entire sequence converges to that limit.

12.2. For the penalty method prove

$$\lim_{k \to \infty} C(x^k, r^k) = f(x^*)$$

12.3. (a) Show that the P in (12.14) has the form required by (12.13).

(b) Assume all g_i, $i = 1, \ldots, m$, have continuous derivatives of order l. Show how to select ϵ so that P also has continuous derivatives of order l.

(c) Assume f and the g_i are concave. Choose $P(x)$ so that $C(x, r)$ is concave.

12.4. For the penalty procedure suppose a point x^1 such that

$$f(x^1) = \max_{x \in E^n} \ f(x)$$

is feasible. Prove that the penalty procedure determines the optimal x^* to problem (12.1) after one iteration.

12.5. Prove Corollary 12.2.1.

12.6. Let $g_i(x)$ be concave. Prove

$$-\frac{1}{g_i(x)}$$

is concave on F^0.

12.7. For the barrier method determine a phase I procedure for calculating x^0 such that

$$g_i(x^0) > 0 \qquad i = 1, \ldots, m$$

12.8. Discuss the duality results for the barrier method using

$$B(x) = -\sum_{i=1}^{m} \frac{1}{g_i(x)^\epsilon}$$

for different ϵ.

12.9. Consider the NLP problem

$$\max \ f(x)$$
$$\text{subject to} \quad g_i(x) \geq 0 \qquad i = 1, \ldots, m^1$$
$$g_i(x) = 0 \qquad m^1 + 1. \ldots, m^2$$

How could the penalty function method be modified to handle this problem? Show that the barrier method cannot solve this problem.

12.10. Prove Lemma 12.2.

12.11. Use the barrier method to solve the example problem given for the penalty method.

12.12. What assumptions are required to assure that the subproblems for both the penalty and barrier methods are solvable by an unconstrained technique? What happens if the unconstrained technique only finds points at which the gradient is zero?

NOTES AND REFERENCES

The penalty method discussion is taken from ZANGWILL (1967a) while discussion of the barrier method can be found in works by FIACCO and by FIACCO and McCORMICK (1964a). These methods, and especially the barrier method, have been examined in detail by Fiacco and McCormick [see FIACCO and McCORMICK (1963), (1964b), (1965a), (1965b), (1966)].

Penalty and barrier methods seem to date back to at least 1943; see COURANT. Other related approaches are found in BELTRAMI and McGILL; BUTLER and MARTIN; CARROLL (1959), (1961); CAMP; FRISCH; GOLDSTEIN and KRIPKE; PIETRZYKOWSKI; POMENTALE; and STONG.

13   Feasible-Direction Methods

and the

Jamming Phenomenon

For the nonlinear programming problem the penalty and barrier methods of Chap. 12 eliminated the constraints by modifying the objective function and solving a sequence of unconstrained problems. Constrained feasible-direction methods instead explicitly consider the constraints and operate as follows: Given a feasible point x^k, a direction d^k is determined. Then x^{k+1} is the feasible point that maximizes the objective function from x^k in the direction d^k.

Unfortunately, the algorithmic map that generates x^{k+1} from x^k need not be closed. This lack of closedness causes onerous complications with feasible-direction methods and can lead to nonconvergence and jamming. When jamming occurs, the algorithm generates a sequence of points $\{x^k\}_1^\infty$ such that the entire sequence converges to a point that is not a solution. Thus the algorithm does not converge. In effect, all the points jam into a "corner." After discussing various suggestions on how to avoid jamming, we prove a convergence theorem that is particularly suited to feasible-direction methods. Finally, we present a feasible-direction method, the ϵ-perturbation method, that by avoiding jamming does converge.

13.1. THE MAP M^3

To properly introduce constrained feasible-direction methods we must first define the map M^3. This map given a feasible point x and direction d determines a point y that optimizes the objective function in the direction d from x. However, in addition, we require that the line segment between x and y be feasible.

For example, it may be that the point

$$x + \tau d$$

is feasible for any τ, $0 \le \tau \le \tau_1$, and any τ, $\tau_2 \le \tau \le \tau_3$, where $0 < \tau_1 < \tau_2 < \tau_3$. Note that between the intervals $[0, \tau_1]$ and $[\tau_2, \tau_3]$ there is a gap and for all τ, $\tau_2 < \tau < \tau_3$, $x + \tau d$ is not feasible. Such a gap can occur if the feasible region is not convex (see exercise 13.11). For this example the map M^3 would optimize only over the first interval $[0, \tau_1]$. Then if $y \in M^3(x, d)$, $y = x + \tau' d$ for some τ', $0 \le \tau' \le \tau_1$. Searching over only the first interval reduces computational effort.

Mathematically, we define M^3 as the set of all y such that

$$f(y) = \max \ \{f(x + \tau d) \,|\, x + \tau d \in F, \tau \in J\}$$

where $y = x + \tau'd \in F$ and for all $0 \le \tau \le \tau'$, $x + \tau d \in F$. Also $J =$ $[0, \alpha]$ for α positive. Thus M^3 only searches over those τ, $0 \le \tau \le \alpha$, for which the line segment x to $x + \tau d$ is feasible.

Constrained Feasible-Direction Methods

Constrained feasible-direction methods employ the map

$$A = M^3 D$$

where given x, the map D determines the direction d.

Observe that this map is somewhat different from the map $A = M^1 D$ employed for the unconstrained feasible-direction methods of Chap. 5, because the constrained case utilizes M^3 instead of M^1 to ensure feasibility.

Regrettably the map M^3 can cause serious difficulties that do not arise for M^1. In Chap. 5 we proved M^1 to be a closed map, and after verifying that the map D was closed, the closedness of $A = M^1 D$ followed easily by the results of Chap. 4, and convergence by Convergence Theorem A was then almost immediate. However, even if D is closed, the map M^3 need not be. Therefore, we cannot prove the map $A = M^3 D$ closed via the results in Chap. 4. Indeed, proof of convergence is often quite difficult.

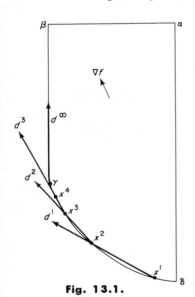

Fig. 13.1.

Examples of M^3 Not Closed†

We now give some examples to illustrate that M^3 need not be closed. In Fig. 13.1 let the feasible region be the area α, β, γ, δ. The straight line segment $[\gamma, \beta]$ is tangent at γ to the arc (γ, δ), which is a portion of a circle.

We choose the directions d^k as follows: Given any point x^k on the arc

---

†This section may be omitted without loss of continuity and the reader may proceed to Sec. 13.2.

(γ, δ), the direction d^k intersects the arc at a point half way between x^k and γ. Figure 13.1 indicates the direction ∇f and the points x^1, x^2, x^3, x^4, etc. Clearly,

$$x^k \longrightarrow \gamma = x^\infty$$

Observe that $d^k \to d^\infty$ where d^∞ is along the line $[\alpha, \beta]$ and that the point β maximizes f along d^∞ from x^∞.

We have constructed a sequence

$$x^{k+1} \in M^3(x^k, d^k)$$

such that

$$x^k \longrightarrow x^\infty = \gamma$$

$$x^{k+1} \longrightarrow x^{\infty+1} = x^\infty = \gamma$$

and

$$d^k \longrightarrow d^\infty$$

However,

$$x^{\infty+1} \notin M^3(x^\infty, d^\infty)$$

Therefore, the map M^3 is not closed at (x^∞, d^∞).

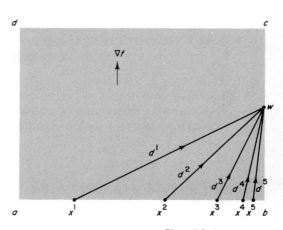

Fig. 13.2.
A linear-constraints example.

A Second Example

One property of the above example is that since the constraint (γ, δ) is curved the direction d^k can cause x^k and x^{k+1} to be on the same constraint. With linear constraints this situation cannot occur; however, even with linear constraints M^3 need not be closed. Consider a second example (see Fig. 13.2), where the constraint region is the rectangle a, b, c, d. A sequence $x^k \longrightarrow b$, the directions $d^k = w - x^k$, and clearly $d^\infty = w - b$.

With ∇f pointing up

$$y^k \in M^3(x^k, d^k)$$

implies

$$y^k = w$$

and

$$\lim y^k = y^\infty = w$$

But $c \in M^3(x^\infty, d^\infty) = M^3(w, w - b)$, and obviously,

$$w \notin M^3(x^\infty, d^\infty)$$

Therefore, M^3 is not closed at (x^∞, d^∞).

These two examples illustrate that the map M^3 need not be closed. Therefore the algorithmic map

$$A = M^3 D$$

of feasible-direction methods with constraints need not be closed, and Convergence Theorems A and B cannot be employed.

13.2. JAMMING IN FEASIBLE-DIRECTION METHODS

As previously discussed, the lack of closedness of M^3 can cause jamming. To motivate the jamming discussion we first prove a lemma and a theorem. In both a feasible-direction algorithm generates the sequence $\{x^k\}_1^\infty$ by use of the algorithmic map

$$A = M^3 D$$

where at each point x^k the map D gives the direction d^k and

$$x^{k+1} = x^k + \tau^k d^k$$

Lemma 13.1 is a pivotal result and we will utilize it not only to further investigate the jamming phenomenon but also to prove the Convergence Theorem for Feasible-Direction Methods. The lemma establishes that a feasible-direction method cannot satisfy three conditions at once.

Lemma 13.1: *Consider a feasible-direction method for the NLP problem where the objective function f is continuously differentiable, and suppose a sequence $\{x^k\}_1^\infty$ is generated.*

Then there cannot be a subsequence $\{x^k\}_{\mathscr{K}}$ that satisfies the following three conditions simultaneously.

(a) $x^k \longrightarrow x^\infty \qquad k \in \mathscr{K}$, *and*

$\qquad d^k \longrightarrow d^\infty \qquad k \in \mathscr{K}$

(b) $\nabla f(x^\infty)^t d^\infty > 0$

and

(c) *A scalar $\delta > 0$ exists such that for all $k \in \mathscr{K}$ and any τ, $0 \leq \tau \leq \delta$, the point*

$$x^k + \tau d^k$$

is feasible.

PROOF: Using (b) we find there is a $\beta > 0$ such that

$$\nabla f(x^\infty)^t d^\infty = \beta > 0 \tag{13.1}$$

Then, because f is continuously differentiable, there must be a $\delta' > 0$ such that any τ, $0 \leq \tau \leq \delta'$, implies

$$\nabla f(x^k + \tau d^k)^t d^k > \frac{\beta}{2} \tag{13.2}$$

if k is sufficiently large and $k \in \mathscr{K}$

Select $\eta > 0$ such that

$$\eta < \quad \min \quad [\delta', \delta, \alpha]$$

where δ is specified in (c) and α is the end point of the interval $J = [0, \alpha]$ used in map M^3. Then from (c)

$$x^k + \eta d^k \quad \text{is feasible} \tag{13.3}$$

Also via choice of x^{k+1} and map M^3

$$f(x^{k+1}) \geq f(x^k + \eta d^k)$$

and utilizing Taylor's expansion, we find

$$f(x^{k+1}) \geq f(x^k) + \eta \nabla f(x')^t d^k$$

where

$$x' = x^k + \theta \eta d^k \quad \text{and} \quad 0 \le \theta \le 1$$

Then (13.2) provides that

$$f(x^{k+1}) \ge f(x^k) + \frac{\eta \beta}{2} \tag{13.4}$$

for k sufficiently large and $k \in \mathcal{K}$

We now call upon Lemma 4.1. Since $x^k \longrightarrow x^\infty, k \in \mathcal{K}$, and the objective function is monotonic

$$\lim_{k \to \infty} f(x^k) = f(x^\infty) \tag{13.5}$$

Consequently, for k sufficiently large, we get

$$f(x^\infty) \le f(x^k) + \frac{\eta \beta}{4} \tag{13.6}$$

Now select $k' \in \mathcal{K}$ so that (13.4) and (13.6) hold. Certainly

$$f(x^{k'+1}) > f(x^\infty) \tag{13.7}$$

However feasible-direction methods are monotonic so that (13.7) cannot hold. Therefore, the three conditions cannot be satisfied simultaneously. ◆

Theorem 13.2, which will now be proved using Lemma 13.1, states the following: suppose a subsequence converges to x^∞ and that for any subsequence converging to x^∞, $x^k \longrightarrow x^\infty$, $k \in \mathcal{K}'$, the corresponding directions have a convergent subsequence $d^k \longrightarrow d^\infty, k \in \mathcal{K}'' \subset \mathcal{K}'$, such that $\nabla f(x^\infty)'d^\infty > 0$. Then under these conditions $x^k \longrightarrow x^\infty, k \longrightarrow \infty$. Basically, the theorem states hypotheses guaranteeing that convergence of a subsequence to a point implies convergence of the entire sequence to that point.

Theorem 13.2: *Consider a feasible-direction method for the NLP problem where the objective function f is continuously differentiable.*

Let there be a subsequence such that

$$x^k \longrightarrow x^\infty \qquad k \in \mathcal{K}$$

Assume also that given any arbitrary subsequence converging to x^∞

$$x^k \longrightarrow x^\infty \qquad k \in \mathcal{K}' \tag{13.8}$$

there is a $\mathcal{K}'' \subset \mathcal{K}'$ such that

$$d^k \longrightarrow d^\infty \qquad k \in \mathcal{K}'' \tag{13.9}$$

where in addition

$$\nabla f(x^\infty)'d^\infty > 0 \tag{13.10}$$

Then under these assumptions, the entire sequence converges to x^∞.

PROOF: By assuming there exists a subsequence that does not converge to x^∞, a contradiction will be established proving the theorem.

Observe that (13.8), (13.9), and (13.10) satisfy (a) and (b) of Lemma 13.1. We will now prove that (c) also holds, thereby providing the contradiction.

Because by hypothesis there is at least one subsequence converging to x^∞, if there is a subsequence not converging to x^∞, there must be some subsequence converging to x^∞

$$x^k \longrightarrow x^\infty \qquad k \in \mathscr{K}'$$

such that

$$\| x^{k+1} - x^k \| > \gamma \tag{13.11}$$

for some $\gamma > 0$. Here, the subsequence $\{x^{k+1}\}_{\mathscr{K}'}$ does not converge to x^∞. By assumption there is a $\mathscr{K}'' \subset \mathscr{K}'$ such that

$$d^k \longrightarrow d'' \qquad k \in \mathscr{K}'' \tag{13.12}$$

We now establish that there must be a $\delta > 0$ such that for k sufficiently large and $k \in \mathscr{K}''$

$$\tau^k > \delta \tag{13.13}$$

For if this statement were not true, there would be a $\mathscr{K}''' \subset \mathscr{K}''$ such that (exercise 13.12)

$$\tau^k \longrightarrow 0 \qquad k \in \mathscr{K}'''$$

Then since $x^{k+1} = x^k + \tau^k d^k$, $\| x^{k+1} - x^k \| = \| \tau^k \delta^k \| \longrightarrow 0$, $k \in \mathscr{K}'''$, which contradicts (13.11). Thus (13.13) must hold.

But since the map M^3 ensures that all points between x^k and x^{k+1} are feasible, (c) of Lemma 13.1 is verified.

In summary, we have exhibited a subsequence for which (a), (b), and (c) of Lemma 13.1 hold simultaneously, and the contradiction is evident. ◆

Interpretation

To interpret the theorem suppose there is a subsequence $\{x^k\}_{\mathscr{K}}$ that converges to x^∞ and that possesses the property

$$\nabla f(x^k)^t d^k \longrightarrow \nabla f(x^\infty)^t d^\infty > 0$$

Clearly d^k provides a good direction to seek an increase in f, but by Lemma 13.1 there cannot be a $\delta > 0$ and a $\mathscr{K}$ such that we can move a distance of δ away from x^k in the direction d^k for $k \in \mathscr{K}$. Hence, it must be that a boundary is nearby and moving in the direction d^k from x^k leads almost immediately into the boundary. It follows, since there does not exist such a $\delta > 0$, that (see exercise 13.12)

$$\tau^k \longrightarrow 0 \qquad k \in \mathscr{K} \tag{13.14}$$

Because of (13.14) the sequence $\{x^{k+1}\}_{\mathscr{K}}$, where

$$x^{k+1} = x^k + \tau^k d^k \tag{13.15}$$

must also converge to x^∞, for taking limits of (13.15) we arrive at

$$\lim_{k \in \mathscr{K}} x^{k+1} = x^\infty + 0 \cdot d^\infty = x^\infty$$

If the directions $\{d^{k+1}\}_{\mathscr{K}}$ for sequence $\{x^{k+1}\}_{\mathscr{K}}$ behave similarly to the directions $\{d^k\}_{\mathscr{K}}$ for $\{x^k\}_{\mathscr{K}}$, then each point x^{k+2} is determined by moving a short distance in the direction d^{k+1} from x^{k+1} and encountering a boundary. Again

$$\tau^{k+1} \longrightarrow 0 \qquad k \in \mathscr{K}$$

and

$$x^{k+2} \longrightarrow x^\infty \qquad k \in \mathscr{K}$$

Thus the subsequence $\{x^{k+2}\}_{\mathscr{K}}$ also converges to x^∞. If we continue in this manner the entire sequence converges to x^∞.

Observe that the entire sequence converged to x^∞ because the directions d^k pointed toward a close boundary. In designing an algorithm, one must ensure that the directions d^k are chosen with adequate regard to the boundaries in the neighborhood of x^k. Specifically, in selecting d^k one must ensure not only that d^k increases the objective function but also that movement in the direction d^k is not obstructed by a constraint. Moreover, this movement must be maintained even in the limit. Thus if $x^k \to x^\infty$ and $d^k \to d^\infty, k \in \mathscr{K}$, and if x^∞ is not a solution, then there must be a $\mathscr{K}^1 \subset \mathscr{K}$ and a $\delta > 0$ such that for any τ, $0 \le \tau \le \delta$, and $k \in \mathscr{K}^1$

$$x^k + \tau d^k \quad \text{is feasible}$$

Figure 13.1 depicts a situation for which this does not occur. Reiterating, suppose $x^k \to x^\infty$, $k \in \mathscr{K}$. Then we want to be sure that there is sufficient room to move along d^k for all $k \in \mathscr{K}$, even in the limit, without bumping into a constraint.

Jamming

The selection of d^k without adequate regard to the boundaries near x^k can lead to nonconvergence. This phenomenon is termed **jamming**, as, by Theorem 13.2, all the points figuratively jam into a corner formed by the boundaries.

Avoiding Jamming

One common error that may lead to jamming is to define d^k by examining only the constraints that are active at x^k. The d^k are thus chosen to yield an increase in f and to avoid only the constraints that are active at x^k. Because any boundaries near x^k are disregarded, the d^k thus chosen might point toward these close boundaries, and jamming could occur (see Fig. 13.3).

In (a) d^k was chosen only to give a good direction with regard to the boundary $g_1(x) = 0$. In (b) a much better d^k was chosen after we observed that although $g_2(x^k) > 0$, $g_2(x^k)$ was sufficiently close to zero so that the boundary $g_2(x) = 0$ was near to x^k. Then d^k was chosen to avoid the boundary $g_2(x)$.

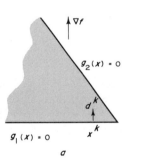
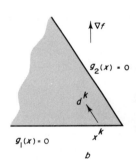

Fig. 13.3.
Choosing d^k.

Another example of how jamming difficulties may arise can be found in the following procedure. Consider the problem

$$\text{max} \quad f(x)$$

$$\text{subject to} \quad x \geq 0$$

where $x \in E^n$ and f is continuously differentiable.

Given any x^k, choose the direction $d^k = d(x^k) = (d(x^k)_i)$ as follows

$$d(x)_i = \begin{cases} 0 & \text{if } \dfrac{\partial f(x)}{\partial x_i} < 0 \quad \text{and} \quad x_i = 0 \\[2mm] \dfrac{\partial f(x)}{\partial x_i} & \text{otherwise} \end{cases}$$

Intuitively, the procedure adjusts x_i^k according to the corresponding component of the gradient $\partial f(x^k)/\partial x_i$. However, if $x_i^k = 0$ and $\partial f(x^k)/\partial x_i < 0$, then changing x_i^k in the direction $\partial f(x^k)/\partial x_i$ would decrease x_i^k, thereby producing an infeasible point. Consequently, in this case $d(x)_i = 0$.

Recall that continuity, or a property similar to it, was useful for convergence. However, observe that the direction function $d(x)$ is discontinuous. For suppose $x^k \longrightarrow x^\infty$, $k \longrightarrow \infty$, where $\partial f(x^\infty)/\partial x_1 < 0$. Also assume

$$x_1^k > 0 \quad \text{for all} \quad k$$

but

$$x_1^\infty = 0$$

Then for all k

$$d(x^k)_1 = \frac{\partial f(x^k)}{\partial x_1}$$

and applying the continuity of ∇f,

$$\lim_{k \to \infty} d(x^k)_1 = \frac{\partial f(x^\infty)}{\partial x_1} < 0$$

But the procedure selects $d(x^\infty)_1 = 0$. The discontinuity is obvious.

Essentially when $x_1^k > 0$ but close to zero, say $x_1^k < \epsilon$, and when $\partial f(x^k)/\partial x_1 < 0$, then in selecting $d(x^k)$ the nearness of the boundary for x_1 was disregarded. The direction $d(x)$ was chosen only by considering the constraints active at x.

That this procedure may not converge has been verified by an example (exercise 13.1) in which jamming occurs and the procedure does not converge.

In addition to considering the boundaries both at and near x, another means of avoiding jamming is by "remembering" the boundaries the procedure has bumped into recently and avoiding them until the points generated are out of that region.

13.3. A CONVERGENCE THEOREM FOR FEASIBLE-DIRECTION METHODS

We now wish to make the above intuitive discussion on how to avoid jamming mathematically precise by stating the Convergence Theorem for Feasible-Direction Methods. The theorem will be proved by direct application of Lemma 13.1, although it can also be established via Convergence Theorem D (exercise 13.10).

Convergence Theorem for Feasible-Direction Methods

Consider a NLP problem whose objective function f is continuously differentiable, and let a solution set be given.

Suppose a feasible-direction algorithm for this problem satisfies the following conditions.

1 *If the algorithm terminates, it terminates at a solution.*
2 *If there exists a convergent subsequence*

$$x^k \longrightarrow x^\infty \qquad k \in \mathcal{K} \tag{13.16}$$

where x^∞ is not a solution, then there is a $\mathcal{K}^1 \subset \mathcal{K}$ such that

(a) $$d^k \longrightarrow d^\infty \qquad k \in \mathcal{K}^1 \tag{13.17}$$

(b) $$\nabla f(x^\infty)^t d^\infty > 0 \tag{13.18}$$

and

(c) *A $\delta > 0$ exists such that for any τ, $0 \leq \tau \leq \delta$,*

$$x^k + \tau d^k \quad \text{is feasible for} \quad k \in \mathcal{K}^1 \tag{13.19}$$

Then an algorithm that satisfies these conditions either terminates at a solution or the limit of any convergence subsequence is a solution.

PROOF: Statement (13.16) and conditions 2(a), 2(b), and 2(c) are a rephrasing of the three conditions of Lemma 13.1. Therefore, a subsequence $\{x^k\}_{\mathcal{K}}$ converging to a point x^∞ that is not a solution cannot exist. ◆

Comments

With many algorithms and particularly feasible-direction methods, it is often easier to prove convergence by assuming that a limit point x^∞ is not a solution and then establishing a contradiction. The Convergence Theorem for Feasible-Direction Methods follows this tack.

Condition 2(a) prohibits such anomalies as $\|d^k\| \longrightarrow \infty$ as $k \longrightarrow \infty$. Usually it is straightforward to ensure this condition holds; simply select all d^k on a compact set. For example, we could let $\|d^k\| \leq 1$ for all k, or calling d_i^k the ith component of d^k, require

$$-1 \leq d_i^k \leq -1 \qquad i = 1, \ldots, n \tag{13.20}$$

for all k.

Condition 2(b) can be interpreted as a condition implied by the fact that x^∞ is not a solution. Then d^∞ is a good direction to increase f.

Finally, 2(c) is the antijamming condition. It guarantees movement in the neighborhood of x^∞ without encountering a boundary.

The Convergence Theorem for Feasible-Direction Methods will be utilized to prove convergence of the next algorithm.

13.4. THE ϵ-PERTURBATION METHOD

To avoid jamming, one approach that was mentioned previously is to determine the directions by examination of all constraints $g_i(x) \leq \epsilon$ for some $\epsilon > 0$. If $g_i(x) = 0$, then x is on that boundary; however, if $0 < g_j(x) < \epsilon$ then x is close to the boundary $g_j(x) = 0$. Examination of all constraints $g_i(x) \leq \epsilon$ will then provide a direction that avoids all boundaries near and at x. The ϵ-perturbation method capitalizes on this technique.

Consider the NLP problem

$$\text{max} \quad f(x)$$
$$\text{subject to} \quad g_i(x) \geq 0 \qquad i = 1, \ldots, m$$

where the f and g_i are continuously differentiable. At any feasible point x define for $\epsilon \geq 0$ the set

$$\mathscr{A}(x, \epsilon) = \{i \,|\, g_i(x) \leq \epsilon\}$$

The indices i in the set $\mathscr{A}(x, \epsilon)$ are from constraints active at x and from constraints whose boundaries are near x. It is this set that will prevent jamming.

The Subproblem

At each iteration the algorithm will solve the following linear programming subproblem given a point x feasible and $\epsilon \geq 0$.

$$\text{max} \quad \sigma$$
$$\nabla g_i(x)^t d - \sigma \geq 0 \qquad i \in \mathscr{A}(x, \epsilon)$$
$$\nabla f(x)^t d - \sigma \geq 0 \qquad\qquad\qquad (13.21)$$
$$1 \geq d_i \geq -1 \qquad i = 1, \ldots, n$$

where here x is fixed and the σ and d are variables, $\sigma \in E^1$ and $d \in E^n$.

For convenience let $(\sigma(x, \epsilon), d(x, \epsilon))$ indicate the optimal point as a function of x and ϵ.

We now examine this problem in some detail. Observe that the point $d = 0$, $\sigma = 0$ is always feasible, therefore,

$$\sigma(x, \epsilon) \geq 0$$

Suppose $\sigma(x, \epsilon) > 0$, then

$$\nabla g_i(x)^t d(x, \epsilon) > 0 \qquad i \in \mathscr{A}(x, \epsilon) \tag{13.22}$$

and

$$\nabla f(x)^t d(x, \epsilon) > 0 \tag{13.23}$$

By (13.23) moving in the direction $d(x, \epsilon)$ will increase f. Also Equation (13.22) provides that a small movement in the direction $d(x, \epsilon)$ will increase g_i for $i \in \mathscr{A}(x, \epsilon)$. In words, by moving a small distance along $d(x, \epsilon)$ we both maintain feasibility and increase f. Consequently, if $\sigma(x, \epsilon) > 0$, the direction $d(x, \epsilon)$ is a good direction in which to proceed.

If $\sigma(x,0) = 0$ for $\epsilon = 0$, the Kuhn-Tucker Conditions Hold

From Chap. 2, if the Kuhn-Tucker (K-T) conditions hold at x

$$\nabla f(x)^t d \leq 0$$

for all directions d such that

$$\nabla g_i(x)^t d \geq 0 \qquad i \in \alpha(x,0) \tag{13.24}$$

Moreover, by scaling the d

$$\nabla f(x)^t d \leq 0$$

for all d such that

$$\nabla g_i(x)^t d \geq 0 \qquad i \in \alpha(x,0)$$

and

$$-1 \leq d_i \leq +1 \tag{13.25}$$

But then (13.25) implies that for $\epsilon = 0$, problem (13.21) has the solution $\sigma(x,0) = 0$.

It has thus been established that if the K-T conditions hold at x,

$$\sigma(x,0) = 0.$$

Regrettably, the converse statement that $\sigma(x,0) = 0$ implies the K-T conditions hold at x is not always true. However, as shown in exercise 13.13, the converse does hold under very mild restrictions. Thus, for most practical applications $\sigma(x,0) = 0$ if and only if the K-T conditions hold at x.

The next two lemmas also clarify the behavior of the subproblem. The first analyzes the set $\mathscr{A}(x, \epsilon)$.

Lemma 13.3: *Suppose*

$$x^k \longrightarrow x^\infty \qquad k \in \mathscr{K}$$

and define

$$\bar{\epsilon} = \tfrac{1}{2}\min\{g_i(x^\infty) \mid g_i(x^\infty) > 0\}$$

where $\bar{\epsilon} = 1$ if all constraints are active at x^∞.
  (a) *If $0 \leq \epsilon \leq \bar{\epsilon}$*

$$\mathscr{A}(x^\infty, \epsilon) = \mathscr{A}(x^\infty, 0)$$

Moreover, for $k \in \mathscr{K}$ sufficiently large

$$\mathscr{A}(x^k, \epsilon) \subset \mathscr{A}(x^\infty, 0) \tag{13.26}$$

for any $\epsilon, 0 \leq \epsilon \leq \bar{\epsilon}$.
  (b) *Given any fixed $\epsilon > 0$, then for k sufficiently large and $k \in \mathscr{K}$*

$$\mathscr{A}(x^k, 0) \subset \mathscr{A}(x^\infty, \epsilon)$$

PROOF of (a): For any $\epsilon \leq \bar{\epsilon}$

$$\mathscr{A}(x^\infty, \epsilon) = \mathscr{A}(x^\infty, 0)$$

because, by definition of $\bar{\epsilon}$, any constraint for which $g_j(x^\infty) \leq \epsilon$ must be such that

$$g_j(x^\infty) = 0$$

For all k sufficiently large and $k \in \mathscr{K}$, via continuity of the constraints

$$g_i(x^k) > \bar{\epsilon} \quad \text{for all} \quad i \quad \text{not in} \quad \mathscr{A}(x^\infty, 0)$$

Therefore, for $\epsilon \leq \bar{\epsilon}$ if i is not in $\mathscr{A}(x^\infty, 0)$, i cannot be in $\mathscr{A}(x^k, \epsilon)$. This proves (a). The proof of (b) is left as exercise 13.5 (see also Fig. 13.4). ◆

The circle around x^k indicates that all constraints whose boundaries are within that circle are included in $\mathscr{A}(x^k, \epsilon)$. Clearly

$$\mathscr{A}(x^\infty, 0) = \{1, 2\}$$

Also

$$\mathscr{A}(x^k, \epsilon) = \{1, 2\}$$

Therefore

$$\mathscr{A}(x^\infty, 0) \subset \mathscr{A}(x^k, \epsilon)$$

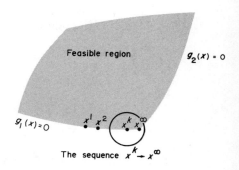

Fig. 13.4.
Geometric illustration of $\mathscr{A}(x^\infty, 0) \subset \mathscr{A}(x^k, \epsilon)$.

The next lemma describes the solutions of the subproblems in more detail.

Lemma 13.4: *Suppose*

$$x^k \longrightarrow x^\infty \qquad k \in \mathscr{K}$$

and

$$\sigma(x^\infty, 0) > 0$$

Then there is an $\bar{\epsilon} > 0$ such that for $k \in \mathscr{K}$ sufficiently large

$$\sigma(x^k, \epsilon) \geq \frac{\sigma(x^\infty, 0)}{2}$$

for any $\epsilon \leq \bar{\epsilon}$.
   In addition, there exists a $\mathscr{K}^1 \subset \mathscr{K}$ such that

$$d(x^k, \epsilon) \longrightarrow d^\infty \qquad k \in \mathscr{K}^1$$

where

$$\nabla f(x^\infty)^t d^\infty > 0$$

PROOF. Define $\bar{\epsilon}$ as in Lemma 13.3, part (a).
We first analyze the problem

$$\begin{aligned}
\max \quad & \sigma \\
& \nabla g_i(x^k)^t d - \sigma \geq 0 \qquad i \in \mathscr{A}(x^\infty, 0) \\
& \nabla f(x^k)^t d - \sigma \geq 0 \\
& 1 \geq d_i \geq -1
\end{aligned} \qquad (13.27)$$

[Observe the use of $\mathscr{A}(x^\infty, 0)$ instead of $\mathscr{A}(x^k, \epsilon)$.]

Since by hypothesis $\sigma(x^\infty, 0) > 0$, the point $\big(d(x^\infty, 0), \sigma'\big)$ where

$$\sigma' = \frac{\sigma(x^\infty, 0)}{2}$$

satisfies

$$\nabla g_i(x^\infty)^t d(x^\infty, 0) - \sigma' > 0 \qquad i \in \mathscr{A}(x^\infty, 0)$$
$$\nabla f(x^\infty)^t d(x^\infty, 0) - \sigma' > 0$$

From continuity of ∇f and ∇g_i, for sufficiently large $k \in \mathscr{K}$

$$\nabla g_i(x^k)^t d(x^\infty, 0) - \sigma' \geq 0 \qquad i \in \mathscr{A}(x^\infty, 0)$$
$$\nabla f(x^k)^t d(x^\infty, 0) - \sigma' \geq 0$$

Therefore $\big(d(x^\infty, 0), \sigma'\big)$ is a feasible point for problem (13.27), and we conclude that the optimal solution of problem (13.27) is at least as great as σ'. Now consider the subproblem for k large and $k \in \mathscr{K}$

$$\begin{aligned} \max \quad & \sigma \\ & \nabla g_i(x^k)^t d - \sigma \geq 0 \qquad i \in \mathscr{A}(x^k, \epsilon) \\ & \nabla f(x^k)^t d - \sigma \geq 0 \\ & 1 \geq d_i \geq -1 \end{aligned} \qquad (13.28)$$

where $\big(\sigma(x^k, \epsilon), d(x^k, \epsilon)\big)$ is the optimal solution.

By Lemma 13.3 part (a), for $\epsilon \leq \bar{\epsilon}$, $\mathscr{A}(x^k, \epsilon) \subset \mathscr{A}(x^\infty, 0)$ so that the subproblem (13.28) is less constrained than subproblem (13.27). Therefore, its solution $\sigma(x^k, \epsilon)$ is at least as large as the solution to (13.27). But the solution to (13.27) is at least σ'. Consequently for $\epsilon \leq \bar{\epsilon}$

$$\sigma(x^k, \epsilon) \geq \frac{\sigma(x^\infty, 0)}{2}$$

for sufficiently large $k \in \mathscr{K}$. This verifies the first part of the theorem.

To conclude the proof examine the constraints in (13.28). Certainly,

$$\nabla f(x^k)^t d(x^k, \epsilon) \geq \sigma' \qquad (13.29)$$

Also since $1 \geq d_i \geq -1$, the $d(x^k, \epsilon)$ are in a compact set. Hence, there is a $\mathscr{K}^1 \subset \mathscr{K}$ such that

$$d(x^k, \epsilon) \longrightarrow d^\infty$$

where by continuity of ∇f and (13.29)

$$\nabla f(x^\infty)^t d^\infty \geq \sigma' \; \blacklozenge$$

Comment

Lemma 13.4 provides the key property required in the hypothesis of Theorem 13.2. Specifically, suppose

$$x^k \longrightarrow x^\infty \qquad k \in \mathcal{K}$$

and

$$d(x^k, \epsilon) \longrightarrow d^\infty \qquad k \in \mathcal{K}$$

If x^∞ does not satisfy the K-T conditions then, as mentioned previously, $\sigma(x^\infty, 0) > 0$, and as shown in Lemma 13.4,

$$\nabla f(x^\infty)^t d^\infty > 0$$

Theorem 13.2 thus concludes that

$$x^k \longrightarrow x^\infty \qquad k \longrightarrow \infty$$

The ϵ-Perturbation Algorithm

We now pose the algorithm. A feasible point x^1 and a scalar $\epsilon^1 > 0$ are given to start the procedure. Then the kth iteration proceeds as follows. The feasible point x^k and $\epsilon^k > 0$ are given. Solve subproblem (13.21) for $\left(\sigma(x^k, \epsilon^k), d(x^k, \epsilon^k)\right)$, where for simplicity call

$$\sigma^k = \sigma(x^k, \epsilon^k) \quad \text{and} \quad d(x^k, \epsilon^k) = d^k$$

If $\sigma^k \geq \epsilon^k$, set

$$\epsilon^{k+1} = \epsilon^k$$

While if $\sigma^k < \epsilon^k$, set

$$\epsilon^{k+1} = \tfrac{1}{2}\epsilon^k$$

Then calculate x^{k+1} from

$$x^{k+1} \in M^3(x^k, d^k)$$

Terminate at x^k if in problem (13.21) $\sigma(x^k, 0) = 0$ for $\epsilon = 0$.

Convergence

We establish convergence using the Convergence Theorem for Feasible-Direction Methods. Recall the f and g_i are assumed continuously differentiable. Let the sequence $\{x^k\}_1^\infty$ define the algorithm. A point x is termed a solution if problem (13.21) yields $\sigma(x,0) = 0$.

Condition 1 of the Convergence Theorem holds easily.

We now establish 2. Let

$$x^k \longrightarrow x^\infty \qquad k \in \mathcal{K}$$

where x^∞ is not a solution. Since $-1 \le d_i \le +1$, all d^k are in a compact set. Hence there is a $\mathcal{K}^1 \subset \mathcal{K}$ such that

$$x^k \longrightarrow x^\infty \qquad k \in \mathcal{K}^1$$

and

$$d^k \longrightarrow d^\infty \qquad k \in \mathcal{K}^1$$

Condition 2(a) is verified.

In order to establish 2(b) and 2(c) we must first examine the sequence $\{\epsilon^k\}_1^\infty$. Clearly, the sequence is nonincreasing and bounded below by zero. It therefore has a limit (exercise 12.1)

$$\lim_{k \to \infty} \epsilon^k = \epsilon^\infty$$

Our first major task is to prove

$$\epsilon^\infty > 0$$

Suppose $\epsilon^\infty = 0$ and we will prove a contradiction. Because x^∞ is not a solution

$$\sigma(x^\infty, 0) > 0 \qquad (13.30)$$

Now consider any arbitrary subsequence converging to x^∞

$$x^k \longrightarrow x^\infty \qquad k \in \mathcal{K}'$$

Using Lemma 13.4 and (13.30) there is a $\mathcal{K}'' \subset \mathcal{K}'$ such that

$$\nabla f(x^\infty)^t d' > 0$$

where

$$d^k \longrightarrow d' \qquad k \in \mathcal{K}''$$

Then employing Theorem 13.2, the entire sequence converges to x^∞

$$\lim_{k \to \infty} x^k = x^\infty \tag{13.31}$$

Let

$$\sigma' = \frac{\sigma(x^\infty, 0)}{2} > 0 \tag{13.32}$$

Applying Lemma 13.4, (13.31), and (13.32)

$$\sigma^k \geq \sigma' > 0 \quad \text{for all} \quad k$$

sufficiently large. It is then impossible for the inequality

$$\sigma^k < \epsilon^k$$

to hold infinitely often. Consequently from construction of ϵ^k in the algorithm, ϵ^{k+1} could not be $\epsilon^k/2$ infinitely often. Therefore $\epsilon^k \longrightarrow \epsilon^\infty = 0$ is impossible and

$$\epsilon^\infty > 0 \tag{13.33}$$

An immediate conclusion from (13.33) is the following: Because $\epsilon^k \longrightarrow \epsilon^\infty > 0$ and by the construction of ϵ^{k+1} from ϵ^k, there must be a P such that for $k \geq P$

$$\epsilon^k = \epsilon^\infty > 0 \tag{13.34}$$

Moreover, since $\sigma^k < \epsilon^k$ implies

$$\epsilon^{k+1} = \tfrac{1}{2}\epsilon^k$$

for all $k \geq P$,

$$\sigma^k \geq \epsilon^\infty > 0 \tag{13.35}$$

Condition 2(b)

Condition 2(b) follows easily from (13.35). Examination of the subproblem constraints reveals that

$$\nabla f(x^k)^t d^k \geq \sigma^k \geq \epsilon^\infty$$

Then by the continuity of ∇f and the fact that $d^k \longrightarrow d^\infty$, $k \in \mathcal{K}^1$,

$$\nabla f(x^\infty)^t d^\infty > 0$$

Condition 2(c)

Lemma 13.3 part (b) and (13.34) provide that for all k sufficiently large and $k \in \mathcal{K}^1$

$$\mathcal{A}(x^\infty, 0) \subset \mathcal{A}(x^k, \epsilon^k)$$

From the solution of the subproblem, since $\sigma^k \geq \epsilon^\infty > 0$,

$$\nabla g_i(x^k)^t d^k \geq \epsilon^\infty \qquad i \in \mathcal{A}(x^\infty, 0) \subset \mathcal{A}(x^k, \epsilon^k)$$

and utilizing continuity not only is

$$\nabla g_i(x^\infty)^t d^\infty \geq \epsilon^\infty > 0 \qquad i \in \mathcal{A}(x^\infty, 0)$$

but also there is a $\delta' > 0$ such that, for $0 \leq \tau \leq \delta'$, and k sufficiently large, $k \in \mathcal{K}^1$,

$$\nabla g_i(x^k + \tau d^k)^t d^k \geq \frac{\epsilon^\infty}{2} \qquad i \in \mathcal{A}(x^\infty, 0)$$

Consequently, since $g_i(x^k) \geq 0$, by Taylor's theorem for $0 \leq \theta \leq 1, 0 \leq \tau \leq \delta'$, and $k \in \mathcal{K}^1$ sufficiently large

$$g_i(x^k + \tau d^k) = g_i(x^k) + \tau \nabla g_i(x^k + \theta \tau d^k)^t d^k$$
$$\geq \tau \frac{\epsilon^\infty}{2} \geq 0 \qquad i \in \mathcal{A}(x^\infty, 0) \tag{13.36}$$

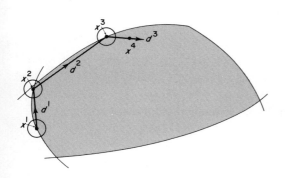

The circles indicate that all constraints whose boundaries are within the circle are included in the index set $\mathcal{A}(x^k, \epsilon)$. At x^2 observe that both g_1 and g_2 are considered in the subproblem. However, x^2 is not on the boundary $g_2(x) = 0$. At x^4 note that the point x^4 is not on any boundary. Hence the operation $M^3(x^3, d^3)$ yields a maximum inside the feasible region.

Fig. 13.5.
Geometric illustration of the ϵ-perturbation procedure.

Interpreting this result we see that if k is sufficiently large and $k \in \mathcal{K}^1$, for any τ such that $0 \leq \tau \leq \delta'$

$$x^k + \tau d^k$$

will not encounter any constraint i for $i \in \mathcal{A}(x^\infty, 0)$. Furthermore, as no other constraints are active at x^∞, there is free movement near x^∞ without bumping into any constraint. In mathematical terms a δ exists, $0 < \delta \leq \delta'$, such that for sufficiently large $k \in \mathcal{K}^1$

$$x^k + \tau d^k \quad \text{is feasible}$$

if $0 \leq \tau \leq \delta$. Condition 2(c) holds and the ϵ-perturbation method converges (see Fig. 13.5).

Comments

The crucial portion of the proof was establishing that $\epsilon^\infty > 0$. Intuitively $\epsilon^\infty > 0$ guarantees that each subproblem considers all the constraints in a neighborhood of x^k (see Fig. 13.5). If $\epsilon^\infty = 0$, then ϵ^k might get so small that $\mathcal{A}(x^k, \epsilon^k)$ would not contain the indices of boundaries at x^∞. All the boundaries at and near the limit point x^∞ must be considered to avoid jamming.

EXERCISES

13.1. Suppose we are given the problem

$$\max \quad f(x)$$
$$x \geq 0$$

where $x \in E^n$ and f is continuously differentiable.

Consider the feasible direction algorithm for this problem

$$A = M^3 D$$

where, given x, the direction $d(x) = (d_i(x))$ is selected as follows

$$d_i(x) = \begin{cases} 0 & \text{if } \dfrac{\partial f(x)}{\partial x_i} < 0 \quad \text{and} \quad x_i = 0 \\[2ex] \dfrac{\partial f(x)}{\partial x_i} & \text{otherwise} \end{cases}$$

Let $f(x) = -\frac{4}{3}[(x_1)^2 - x_1 x_2 - (x_2)^2]^{3/4} + x_3$ for $x \in E^3$.
  (a) Prove that ∇f is continuous.
  (b) f is concave.
  Show that if we start from $(x_1, x_2, x_3) = (1, 0, \frac{1}{2})$, the sequence x^k converges to a point which is not optimal.
  (c) Prove that f does not have continuous second partial derivatives.

13.2. Use the ϵ-perturbation method on the problem

$$\max \quad f(x)$$

$$\text{subject to} \quad x \geq 0$$

for $x \in E^3$ where

$$f(x) = -\frac{4}{3}[(x_1)^2 - x_1 x_2 - (x_2)^2]^{3/4} + x_3$$

How does this method avoid jamming?

13.3. Suppose the NLP problem has the form

$$\max \quad f(x)$$

$$\text{subject to} \quad g_i(x) \geq 0 \qquad i = 1, \ldots, m^1$$

$$a_i^t x \geq b_i \qquad i = m^1 + 1, \ldots, m$$

Thus the constraints $i = m^1 + 1, \ldots, m$ are linear.
  Define

$$\mathscr{A}(x, \epsilon) = \{i \mid g_i(x) \leq \epsilon\}$$

$$\mathscr{B}(x, \epsilon) = \{i \mid a_i^t x - b_i \leq \epsilon\}$$

  Prove that ϵ-perturbation method still converges by using the subproblem

$$\max \quad \sigma$$

$$\nabla f(x^k)^t d - \sigma \geq 0$$

$$\nabla g_i(x^k)^t d - \sigma \geq 0 \qquad i \in \mathscr{A}(x^k, \epsilon)$$

$$a_i^t d \geq 0 \qquad i \in \mathscr{B}(x^k, \epsilon)$$

$$-1 \leq d_i \leq +1$$

13.4. Suppose in the ϵ-perturbation procedure that the subproblems were

$$\max \quad f(x^k)^t d$$

$$\nabla g_i(x^k)^t d \geq 0 \qquad i \in \mathscr{A}(x^k, 0)$$

$$-1 \leq d_i \leq 1$$

where x^k is fixed. The solution d^k would be the direction at the kth iteration. Prove that this direction may not be a feasible direction. That is, for all $\tau > 0$, $x^k + \tau d^k$ might be infeasible. Therefore, this subproblem cannot be used.

13.5. Prove part (b) of Lemma 13.3.

13.6. The unconstrained methods of Chap. 5 are also feasible-direction methods, and because the feasible region is the entire space, the maps M^1 and M^3 become identical. Prove convergence of the algorithms in Chap. 5 using the Convergence Theorem for Feasible-Direction Methods.

13.7. Utilize the Convergence Theorem for Feasible-Direction Methods to prove convergence of the linear-approximation method of Chap. 9. How is jamming avoided?

    Hint: Each subproblem considers all the constraints.

13.8. Suppose we must solve the problem

$$\max\ f(x)$$
$$\text{subject to}\quad g_i(x) \geq 0 \qquad i = 1, \ldots, m^1$$
$$a_i^t x \geq b_i \qquad i = m^1 + 1, \ldots, m$$

where the f and g_i are continuously differentiable and the constraints $i = m^1 + 1, \ldots, m$ are linear.

    Consider a feasible-direction method for this problem where the directions are calculated in the following manner:

    Let $\epsilon > 0$ be given, and define

$$\mathcal{B}(x) = \{i \mid a_i^t x - b_i \leq \epsilon\}$$

Also let

$$h(d, x) = \quad \min\ [\nabla f(x)^t d, g_1(x) + \nabla g_1(x)^t d,$$
$$g_2(x) + \nabla g_2(x)^t d, \ldots, g_{m^1}(x) + \nabla g_{m^1}(x)^t d]$$

    Fix $\epsilon > 0$ throughout the algorithm. For x^k feasible calculate the direction d^k to solve

$$\max\ h(d, x^k)$$
$$a_i^t d + a_i^t x^k - b_i \geq 0 \qquad i \in \mathcal{B}(x^k)$$
$$-1 \leq d_i \leq +1 \qquad i = 1, \ldots, n$$

Then $x^{k+1} \in M^3(x^k, d^k)$.

    Prove convergence of this procedure.

13.9. *A Modification of the Convex-Simplex Method.* Consider an algorithm precisely the same as the convex-simplex method except for the tableau-selection technique. Instead of the tableau-selection technique given in Chap. 9 use the following criterion.

Suppose x_s is the variable being either increased or decreased. If by adjusting x_s a basic variable, say x_{B_r}, becomes zero, pivot on $t_{r,s}$, placing x_s in the basis and removing x_{B_r} from the basis. If no basic variable becomes zero, do not pivot.

Design an appropriate nondegeneracy assumption and prove convergence of this method.

Show that the CSM-CD method of Chap. 9 converges in a finite number of steps for a quadratic using this new tableau-selection rule.

13.10. Prove the Convergence Theorem for Feasible-Direction Methods by establishing that it is a special case of Convergence Theorem D.

13.11. Consider the map M^2 defined in Chap. 9.
  (a) Show graphically that in calculating the map M^2 the optimization can be over several disconnected intervals.
  (b) Prove that M^2 and M^3 are identical if the feasible region is convex and $J = [0, a]$.
  (c) Show that M^2 need not be a closed map.

13.12. Consider a sequence $\{\tau^k\}_1^\infty$ where $\tau^k \geq 0$ for all k. Prove if there does not exist a $\delta > 0$ such that

$$\tau^k > \delta \qquad k \in \mathcal{K}''$$

then there is a $\mathcal{K}''' \subset \mathcal{K}''$ such that

$$\tau^k \longrightarrow 0 \qquad k \in \mathcal{K}'''$$

13.13. Suppose subproblem (13.21) has a solution $\sigma(x,0) = 0$.
  (a) Show that $\sigma(x,0) = 0$ is also the solution to the same subproblem but with the constraints $-1 \leq d_i \leq +1$ deleted.
  (b) Use the dual theorem of linear programming to prove that
$$\lambda_i g_i(x) = 0 \qquad i = 1, \ldots, m$$
  and
$$\lambda_0 \nabla f(x) + \sum_{i=1}^m \lambda_i \nabla g_i(x) = 0$$
  for $\lambda_i \geq 0, i = 0, \ldots, m$.
  (c) Prove that if the vectors $\nabla g_i(x)$, $i \in \alpha(x,0)$, are linearly independent, then the K-T conditions hold at x.
  (d) State other conditions that imply the K-T conditions hold at x.

NOTES AND REFERENCES

§13.1. and **13.2.** Jamming is used here, although the term zigzagging has been previously suggested by ZOUTENDIJK (1960). The term jamming seems more descriptive, in that it implies that something does not work—it is jammed, or in NLP phraseology, it does not converge. Many algorithms zigzag in the sense of geometrically describing a back-and-forth path, yet still converge. For the earliest discussions of jamming, see the works of ZOUTENDIJK (1960) and ROSEN (1961). Refer also to WOLFE (1966) and ZANGWILL (1967h).

That the map M^3 not being closed may cause jamming is a new observation. The example of a procedure that actually does jam was suggested by WOLFE (1966). Exercise 13.1 is based on his example. For another example stated in geometric terms, see the paper by ZANGWILL (1967h).

§13.3. For a theorem related to, although somewhat different from, the Convergence Theorem for Feasible-Direction Methods, refer to TOPKIS and VEINOTT. Exercise 13.10 is also based on TOPKIS and VEINOTT.

§13.4. This method was suggested by the work of ZOUTENDIJK (1960), although the proof is new.

14   Cutting-Plane Algorithms

This chapter presents three cutting-plane algorithms, the concave-cutting-plane method, the supporting-hyperplane method, and the dual-cutting-plane method. Cutting-plane algorithms are interesting because the algorithmic map takes a set, "cuts" a portion off the set, and thereby forms a new set. The points z on which the maps operate are thus sets in E^n. After introducing the general theory of these algorithms we present a new convergence theorem particularly suited to cutting-plane methods and prove convergence of the three algorithms.

14.1. THE THEORY OF CUTTING-PLANE ALGORITHMS

Cutting-plane algorithms operate on points z that are actually sets in E^{n_1}. Given a set z^k, the algorithm determines a half space H^k; then the successor

$$z^{k+1} = z^k \cap H^k \tag{14.1}$$

The goal of these algorithms is to calculate a point in some specified set G, where G is either the feasible region or some other set related to the nonlinear programming (NLP) problem. But because the set G is often difficult to handle, the algorithms, instead of attacking G directly, start with a simpler set z^1 that approximates G. Using z^1 and a special map Γ, to be discussed subsequently, we calculate a test point

$$w^1 \in \Gamma(z^1) \tag{14.2}$$

such that

$$w^1 \in z^1 \tag{14.3}$$

The point w^1 is given a certain test, called a solution test, to determine if it is in G. If it passes the test, the procedure terminates for we will have located a point in G.

If w^1 does not pass the solution test, we employ a second map $\Delta: E^{n_1} \longrightarrow E^{n_2}$ to calculate a point

$$v^1 \in \Delta(w^1)$$

Using v^1 we construct a half space $H(v^1) \subset E^{n_1}$. Then

$$z^2 = z^1 \cap H(v^2)$$

The half space has the property that

$$w^1 \notin H(v^1)$$

and consequently, from (14.3)

$$z^2 \subset\subset z^1$$

where $\subset\subset$ indicates strict containment.

The set z^2 approximates G more closely than z^1, and with this better approximation we hope to determine a point in G. By use of z^2, a point

$$w^2 \in \Gamma(z^2)$$

is calculated. If w^2 passes the solution test, we terminate. Otherwise the procedure continues in the same manner.

In general, given z^k, a test point $w^k \in \Gamma(z^k)$ is calculated. If w^k passes the solution test, we terminate; if not, we determine a point $v^k \in \Delta(w^k)$ and corresponding set $H(v^k)$. Then

$$z^{k+1} = z^k \cap H(v^k)$$

Clearly

$$z^{k+1} \subset z^k \subset z^1 \quad \text{for all} \quad k \tag{14.4}$$

Also note that although $w^k \notin H(v^k)$ the algorithm does generate

$$w^l \in z^l$$

It therefore follows from (14.4) that

$$w^l \in z^{k+1} \quad \text{for all} \quad l \geq k + 1 \tag{14.5}$$

In general the sets $H(v)$ have the form

$$H(v) = \{x \,|\, a(v) + b(v)^t x \geq 0\} \tag{14.6}$$

where a and b are functions $a: E^{n_2} \longrightarrow E^1$ and $b: E^{n_2} \longrightarrow E^{n_1}$.

We can now specify the algorithmic map A for cutting-plane methods precisely.

Operation of a Cutting-Plane Method

For a given algorithm a solution test for $w \in \Gamma(z)$ will be given, which is constructed to ensure that any w that passes the test solves the NLP problem.

Now suppose, given z, we desire to calculate a successor point $y \in A(z)$. First via the map Γ calculate the test point

$$w \in \Gamma(z) \tag{14.7}$$

If w passes the solution test, terminate the procedure.

Otherwise, using the map $\Delta: E^{n_1} \rightarrow E^{n_2}$, we determine

$$v \in \Delta(w) \tag{14.8}$$

and define a set $H(v) \subset E^{n_1}$ where

$$H(v) = \{x \mid a(v) + b(v)x^t \geq 0\}$$

for functions $a: E^{n_2} \rightarrow E^1$ and $b: E^{n_2} \rightarrow E^{n_1}$. Then, we formulate $y \in A(z)$ by

$$y = z \cap H(v)$$

To initiate the procedure, a set z^1 is given.

Briefly, given z^k, calculate $w^k \in \Gamma(z^k)$. If $w^k \notin G$, then specify the set $H(v^k)$ using $v^k \in \Delta(w^k)$, and $z^{k+1} = z^k \cap H(v^k)$. In this manner we generate a sequence $\{z^k\}_1^\infty$. (It should be observed that $b(v) \neq 0$ was not required. Therefore, the set $H(v)$ is not necessarily a bona fide half space, although it generally is. In addition, the map $\Delta(w)$ need only be defined for w that do not pass the solution test.)

The convergence theorem for cutting-plane methods can now be established. We will only consider the case for which an infinite sequence $\{z^k\}_1^\infty$ is generated, as the case of finite termination is trivial. The notation is the same as specified above.

Cutting-Plane Convergence Theorem: *Let a cutting-plane algorithm generate a sequence of sets* $\{z^k\}_1^\infty$ *and corresponding sequences of points* $\{w^k\}_1^\infty$ *and* $\{v^k\}_1^\infty$. *Suppose*

(1) *All points w^k are on a compact set, and all points v^k are contained in a compact set.*

(2) *For any z, $w \in \Gamma(z)$ implies*

$$w \in z$$

(3) *The map $\Delta(w)$ is closed for any w that does not pass the solution test. Also the functions a and b are continuous.*

(4) *If w does not pass the solution test, then for any $v \in \Delta(w)$,*

$$w \notin H(v) = \{x \mid a(v) + b(v)^t x \geq 0\}$$

and

$$z \cap H(v) \neq \phi$$

Then if the algorithm satisfies these four conditions, for some $\mathscr{K}$

$$w^k \longrightarrow w^\infty \qquad k \in \mathscr{K}$$

where w^∞ passes the solution test

PROOF: Observe using condition (4) that all z^k exist and are not null. Condition 1 ensures that there must be a $\mathscr{K}$ for which

$$w^k \longrightarrow w^\infty \qquad k \in \mathscr{K} \tag{14.9}$$

and

$$v^k \longrightarrow v^\infty \qquad k \in \mathscr{K} \tag{14.10}$$

Using condition 2, we find [see (14.5)]

$$w^l \in H(v^k) \quad \text{for all} \quad l \geq k+1$$

or explicitly

$$a(v^k) + b(v^k)^t w^l \geq 0 \quad \text{for all} \quad l \geq k+1$$

Then from (14.9)

$$a(v^k) + b(v^k)^t w^\infty \geq 0$$

and calling upon condition 3 and (14.10), we find

$$a(v^\infty) + b(v^\infty)^t w^\infty \geq 0 \tag{14.11}$$

or equivalently

$$w^\infty \in H(v^\infty) \tag{14.12}$$

Now suppose w^∞ does not pass the solution test. Applying condition 3 because Δ is a closed map

$$v^\infty \in \Delta(w^\infty)$$

But, for such a w^∞ condition 4 guarantees,

$$w^\infty \notin H(v^\infty) \tag{14.13}$$

As (14.12) and (14.13) are contradictory, w^∞ must pass the solution test. ◆

Comments

Condition 1 ensures that the subsequences are well-behaved. Conditions 2 and 4 combine to provide that $H(v^k)$ actually "cuts off" w^k, thereby obtaining a strict improvement with

$$z^{k+1} \subset\subset z^k$$

However, the second part of 4 ensures that the half space $H(v^k)$ does not cut off too much leaving nothing. Condition 3 specifies the continuity and closedness conditions that, as previously mentioned, are helpful for convergence.

The Map Γ is a Linear Programming Problem

We begin our study of the three cutting-plane algorithms by investigating the map Γ, which has the same form for each method. In particular, given any Euclidean set z, Γ yields an optimal solution to the subproblem

$$\max \quad q^t x$$
$$x \in z \tag{14.14}$$

More precisely,

$$\Gamma(z) = \{x^* \,|\, x^* \text{ is optimal for problem (14.14)}\}$$

For each of the methods a maximining point will exist.

The set z^1 in all three methods has the form

$$z^1 = \{x \,|\, Ax \geq b\}$$

Consequently, finding

$$w^1 \in \Gamma(z^1)$$

is a linear programming (LP) problem. Moreover, since

$$z^2 = z^1 \cap H(w^1)$$

it follows from (14.6) that finding

$$w^2 \in \Gamma(z^2)$$

is a LP problem. Clearly, the problem of finding $w^k \in \Gamma(z^k)$ is also a LP problem. Thus the three cutting-plane methods solve the NLP problem by instead solving a sequence of LP problems.

14.2. THE CONCAVE-CUTTING-PLANE ALGORITHM

The concave method is for the NLP problem when f and the g_i are all concave.

Transforming the Problem

Consider problem (14.15)

$$\max \quad f(x)$$
$$\text{subject to} \quad g_i(x) \geq 0 \qquad i = 1, \ldots, m \tag{14.15}$$

where $x \in E^n$.

Defining a scalar w, we find problem (14.15) is equivalent to

$$\max \quad w$$
$$\text{subject to} \quad f(x) - w \geq 0 \tag{14.16}$$
$$g_i(x) \geq 0 \qquad i = 1, \ldots, m$$

In effect, by the addition of one constraint and one variable, we have changed problem (14.15) into a problem with a linear objective function.

Let

$$g(x, w) = \quad \min \quad [f(x) - w, g_i(x), \ldots, g_m(x)]$$

Then problem (14.16) is equivalent to

$$\max \quad w$$
$$\text{subject to} \quad g(x, w) \geq 0$$

Consequently, we have reduced problem (14.15) to a problem with one constraint and a linear objective function. Furthermore, the function $g(x, w)$ is concave by property (e) of Chap. 2.

Changing notation somewhat, we see that it is sufficient to focus on the problem

$$\max \quad q^t x$$
$$\text{subject to} \quad g(x) \geq 0 \tag{14.17}$$

where g is a single concave function. Clearly problem (14.17) is equivalent to problem (14.15). The concave-cutting-plane method is designed to solve problem (14.17).

Assumptions

The concave algorithm for problem (14.17) assumes that the feasible set F for problem (14.17) is contained in a compact set U where

$$U = \{x \mid Ax \leq b\} \subset E^n \tag{14.18}$$

We also suppose that the constraint function g is continuous. Finally, assume that given any w, there is a vector $u(w)$, which is uniformly bounded on U, such that

$$g(x) \leq g(w) + u(w)^t(x - w) \tag{14.19}$$

for all $x \in U$. By *uniformly bounded* is meant that there is a positive number M such that $\|u(w)\| \leq M$ for all $w \in U$.

We may now completely define the maps which comprise the concave-cutting-plane method.

Specification of the Algorithm

Given a point z, define the subproblem

$$\begin{array}{c} \text{map}\quad q^t x \\ x \in z \end{array} \tag{14.20}$$

Then the map Γ is expressed as

$$\Gamma(z) = \{x^* \mid x^* \text{ is optimal for problem (14.20)}\}$$

Given a point $w \in \Gamma(z)$, the solution test is to see if

$$w \in F \tag{14.21}$$

where F is the feasible set for problem (14.17).

The map Δ is defined in terms of w and of the $u(w)$ in Equation (14.19),

$$\Delta(w) = \{(w, u(w))\} \tag{14.22}$$

Using the previous notation, we find if $v \in \Delta(w)$, $v = (w, u(w))$. Finally, for $(w, u(w)) \in \Delta(w)$

$$H(v) = \{x \mid g(w) + u(w)^t(x - w) \geq 0\}$$

To begin the procedure, let

$$z^1 = U$$

The algorithm is specified.

Essentially, the procedure operates as follows: Given z^k, let w^k solve problem (14.20). Should $w^k \in F$ stop; otherwise, after determining $u(w^k)$ as in (14.19), form

$$H(v^k) = \{x \mid g(w^k) + u(w^k)^t(x - w^k) \geq 0\}$$

and set

$$z^{k+1} = z^k \cap H(v^k)$$

As mentioned in Sec. 14.1, all subproblems will be LP problems.

A method for selecting $u(w)$ for problem (14.17) such that Δ is closed for all $w \in U$ is left as exercise 14.1. Moreover, that exercise also indicates that the $(w^k, u(w^k))$ are in a compact set.

Before proving convergence, we make the following observation.

Lemma 14.1: *For all k*

$$F \subset z^k$$

PROOF: Let $x \in F$, then

$$0 \leq g(x)$$

Also from (14.19)

$$0 \leq g(x) \leq g(w^k) + u(w^k)^t(x - w^k)$$

Consequently

$$x \in H(w^k) \quad \text{for all} \quad k$$

so that

$$F \subset H(w^k) \quad \text{for all} \quad k$$

But since

$$z^k = z^1 \bigcap_{l=1}^{k-1} H(w^l)$$

and by assumption $F \subset U = z^1$

$$F \subset z^k \quad \text{for all} \quad k \quad \blacklozenge$$

Now convergence will be established via the Cutting-Plane Convergence Theorem.

Condition 1: For all k, $w^k \in z^1 = U$, which is by assumption compact. Also all $v^k = (w^k, u(w^k))$ are in a compact set, via exercise 14.2.

Condition 2: This is obvious by the definition of Γ.

Condition 3: The map Δ is closed by exercise 14.2. Clearly $a(v^k) = g(w^k) - u(w^k)^t w^k$ and $b(v^k) = u(w^k)$ are continuous because g is continuous and $v^k = (w^k, u(w^k))$.

Condition 4: If w does not pass the solution test

$$w \notin F$$

so that

$$g(w) < 0$$

but then

$$w \notin H(v) = \{x \mid g(w) + u(w)^t(x - w) \geq 0\}$$

In addition, Lemma 14.1 ensures

$$F \subset z^k \cap H(v^k)$$

The algorithm converges.

Interpretation of Convergence

Let x^* be optimal for problem (14.17). By Lemma 14.1 and problem (14.20) it must be that

$$q^t x^* \leq q^t w^k \quad \text{for all} \quad k$$

Consequently, when $w^k \in F$, w^k must be optimal for problem (14.17). The concave method therefore generates an optimal point to problem (14.17).

14.2.1. The Operation of the Concave Method in E^1

Let $q \cdot x$ and $g(x)$ be represented as in Fig. 14.1, and suppose

$$U = \{x \mid a \leq x \leq b\}$$

The feasible region is $F = \{x \mid \delta \leq x \leq \epsilon\}$.

As $z^1 = U$, the first problem is

$$\max \quad qx$$
$$a \leq x \leq b$$

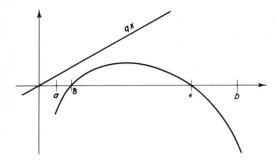

Fig. 14.1.

with optimal solution $w^1 = b$.

Figure 14.2 indicates $g(w^1) + u(w^1)^t(x - w^1)$. Observe that as $H(v^1) = \{x \mid x \le c\}$, $z^2 = \{x \mid a \le x \le b \text{ and } x \le c\}$. The second subproblem becomes

$$\max \quad qx$$
$$a \le x \le b$$
$$x \le c$$

with an optimal solution $w^2 = c$.

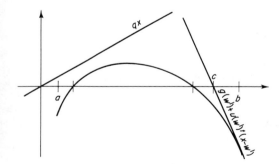

Fig. 14.2.

Figure 14.3 shows $z^3 = \{x \mid a \le x \le b, x \le c, x \le d\}$, and the optimal solution to the third subproblem is $w^3 = d$. The procedure continues in this manner.

From a geometric viewpoint, the hyperplanes

$$g(w^k) + u(w^k)^t(x - w^k)$$

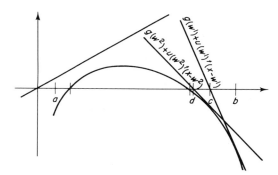

Fig. 14.3.

are used to cut off a portion of the set. Hence the name cutting-plane algorithm.

   The concave method depends crucially on the concavity of the f and the g_i (exercise 14.3). Indeed, the algorithm may not work even if the feasible region is convex but the constraints are not concave functions. However, the following algorithm overcomes this limitation.

14.3. THE SUPPORTING-HYPERPLANE ALGORITHM

   Consider problem (14.15) where f is concave and the g_i are quasi-concave. As discussed with the concave method we may transform the problem into one with a linear objective function. Consider, therefore, the problem

$$\max \quad q^t x$$
$$\text{subject to} \quad g_i(x) \geq 0 \qquad i = 1, \ldots, m \tag{14.23}$$

where the g_i are quasi-concave and continuously differentiable (exercise 14.4). Observe that by quasi-concavity, the feasible region of problem (14.23) is a convex set.

   The supporting-hyperplane algorithm is designed to solve problem (14.23). It is similar to the concave method except for the map Δ and the cutting half space $H(v)$. The half space will be constructed on the boundary of the convex feasible region F, where a point u is on the boundary of F if

$$g_j(u) = 0 \quad \text{for some} \quad j$$

and

$$g_i(u) \geq 0 \quad \text{all} \quad i$$

Assumptions

The algorithm assumes that all functions are continuously differentiable and that the convex feasible set F is contained in a compact set U

$$U = \{x \mid Ax \leq b\} \tag{14.24}$$

Furthermore, it is supposed that there is a point a such that $g_i(a) > 0$ for all i and that if U is on the boundary and $g_i(u) = 0$, then $\nabla g_i(u) \neq 0$.

Specification of the Algorithm

The algorithm is identical to the concave method except for the map Δ and half space $H(v)$. Clearly, we need only specify $\Delta(w)$ for w that do not pass the solution test. Given such a w, because the solution test is the same as the concave method,

$$w \notin F$$

Using the point a inside F let u be a point on the line between a and w such that u is on the boundary of F (exercise 14.6). This u is what Δ determines. Specifically, given $w \notin F$

$$\Delta(w) = \{u\}$$

where u is on the boundary between a and w.

Using $u \in \Delta(w)$ we express the set $H(u)$ for any j such that $g_j(u) = 0$ as

$$H(u) = \{x \mid \nabla g_j(u)^t(x - u) \geq 0\}$$

It follows from convexity of F that (exercise 14.5)

$$F \subset H(u) \tag{14.25}$$

The similarity between this procedure and the concave method should be clear. Both start with a point z and solve the subproblem for w. If w is feasible, both terminate. Should w not be feasible, then unlike the concave method, the hyperplane approach calculates a boundary point u on the line between the interior point a and the infeasible w. At u the half space $H(u)$ is constructed, and the successor point is then the intersection of $H(u)$ and z.

Convergence Proof

Condition 1: All $w^k \in U$, which by assumption is compact. Also $v^k \in \Delta(w^k)$ implies $v^k = u^k$, but all u^k are in U.

Condition 2: This is obvious.

Condition 3: We now must prove

$$\Delta(w) = \{u\}$$

is closed. Let

$$w^k \longrightarrow w^\infty \qquad k \in \mathcal{K}$$

and

$$u^k \longrightarrow u^\infty \qquad k \in \mathcal{K}$$

Because u^k is on the line between a and w^k, there exists θ^k such that

$$u^k = \theta^k w^k + (1 - \theta^k)a \qquad 0 \le \theta^k \le 1$$

Since the $0 \le \theta^k \le 1$, there is a $\mathcal{K}^1 \subset \mathcal{K}$ such that

$$\theta^k \longrightarrow \theta^\infty$$

Consequently

$$u^\infty = \theta^\infty w^\infty + (1 - \theta^\infty)a$$

and u^∞ is on the line between w^∞ and a.

It now must be verified that u^∞ is a boundary point. Because there are only a finite number of constraints, there must be a j and a $\mathcal{K}^2 \subset \mathcal{K}^1$ such that

$$g_j(u^k) = 0 \qquad k \in \mathcal{K}^2$$

and

$$g_i(u^k) \ge 0 \qquad k \in \mathcal{K}^2 \quad \text{for all} \quad i$$

By continuity

$$g_j(u^\infty) = 0$$
$$g_i(u^\infty) \ge 0 \quad \text{for all} \quad i$$

and u^∞ is on the boundary.

Therefore because u^∞ is a boundary point between a and w^∞,

$$u^\infty \in \Delta(w^\infty)$$

and the map Δ is closed.

Clearly $a(v^k) = -\nabla g_j(u^k)^t u^k$ and $b(v^k) = \nabla g_j(u^k)$ are continuous because $u^k = v^k$.

Condition 4: (exercise 14.7). If $w \notin F$,

$$w \notin H(u) = \{x \mid \nabla g_j(u)^t(x - u) \geq 0\}$$

Furthermore by (14.25)

$$F \subset z^k \cap H(u^k)$$

The method converges.

Interpretation of Convergence

Via (14.25) it is known, as in the concave method, that

$$F \subset z^k \quad \text{for all} \quad k$$

Therefore, if w passes the solution test, w is optimal for problem (14.23). The supporting hyperplane thereby generates an optimal point for problem (14.23).

14.3.1. The Geometry of the Supporting-Hyperplane Algorithm in E^2

Let the rectangle α, β, γ, δ (see Fig. 14.4) be the initial region $U = z^1$. The feasible set is F, the arrow indicates the direction that maximizes the objective function $q^t x$, and the point a is indicated. The solution to the first subproblem has $w^1 = \gamma$.

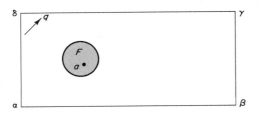

Fig. 14.4.

The point u^1 on the line between a and γ is shown in Fig.14.5. Observe how the hyperplane $\nabla g_{i_1}(u^1)^t(x - u^1)$ touches or supports F at u^1. The new set z^2 is the quadrilateral α, ϵ, η, δ; the solution to the second subproblem would be the point η.

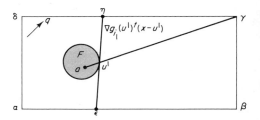

Fig. 14.5.

The point u^2 is indicated in Fig. 14.6. The set z^3 is the five-sided figure α, ϵ, τ, σ, δ, and the solution to the third subproblem is at σ. The procedure continues in this manner.

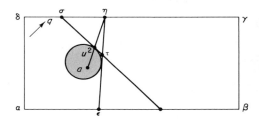

Fig. 14.6.

14.4. A DUAL-CUTTING-PLANE ALGORITHM

Given problem (14.15) or a modification thereof, the previously discussed cutting-plane methods attempted to approximate the feasible region F. The dual-cutting-plane algorithm instead exploits the Lagrangean and the dual-programming problem.

Lagrangean Results

Recall from Chap. 2 the Lagrangean

$$L(x, \lambda) = f(x) + \sum_{i=1}^{m} \lambda_i g_i(x) = f(x) + \lambda^t g(x)$$

where $g(x) = (g_i(x))$, the primal function

$$L_*(x) = \min_{\lambda \geq 0} \; L(x, \lambda)$$

and the dual function

$$L^*(\lambda) = \max_{x \in X} \; L(x, \lambda)$$

It was proved in Chap. 2 that

$$L_*(x) \leq L^*(\lambda) \tag{14.26}$$

The primal problem of calculating

$$\max_{x \in X} \; L_*(x)$$

was equivalent to

$$\max \; f(x)$$
$$\text{subject to} \quad g_i(x) \geq 0 \qquad i = 1, \ldots, m \tag{14.27}$$
$$x \in X$$

The dual problem of determining

$$\min_{\lambda \geq 0} \; L^*(\lambda)$$

could also be written as

$$\min \; f(x) + \lambda^t g(x)$$
$$\text{subject to} \quad f(x) + \lambda^t g(x) = \max_{x \in X} \; \{f(x) + \lambda^t g(x)\} \tag{14.28}$$
$$\lambda \geq 0$$

Should $(\bar{x}, \bar{\lambda})$ be a saddle point to the Langrangean, then it was also proved that $\bar{x}$ is optimal for problem (14.27) and $(\bar{x}, \bar{\lambda})$ is optimal for problem (14.28).

The Dual-Cutting-Plane Assumptions

The dual-cutting-plane algorithm proceeds by approximating problem (14.28). It assumes that the objective function and all constraints are continuous and concave and that there is a point

$$x^0 \in X$$

for which

$$g_i(x^0) > 0 \qquad i = 1, \ldots, m \qquad (14.29)$$

Also, the set X is assumed convex and compact in E^n.

The sets z^k will be sets in E^{n+1} of the form

$$z^k = \left\{ (\lambda, \mu) \middle| \begin{matrix} f(x^i) + \lambda^t g(x^i) - \mu \le 0 & i = 0, \ldots, k-1 \\ \lambda \ge 0 \end{matrix} \right\} \qquad (14.30)$$

where $\lambda \quad E^n$, $\mu \quad E^1$, and the points x^i, $i = 0, \ldots, k-1$, are fixed. We define the initial set

$$z^1 = \{ (\lambda, \mu) \,|\, f(x^0) + \lambda^t g(x^0) - \mu \le 0, \ \lambda \ge 0 \} \qquad (14.31)$$

where x^0 is given in (14.29).

The Algorithmic Map

The algorithmic map is as follows. Initiate the procedure with the z^1 in (14.31). In general given z^k, let (λ^k, μ^k) be the optimal solution to the LP subproblem

$$\min \{ \mu \,|\, (\lambda, \mu) \in z^k \}$$

or explicitly

$$\min \quad \mu$$
$$\text{subject to} \quad f(x^i) + \lambda^t g(x^i) - \mu \le 0 \qquad i = 0, \ldots, k-1 \quad (14.32)$$
$$\lambda \ge 0$$

In the previous notation $w^k \in \Gamma(z^k)$ implies that w^k is an optimal solution to (14.32); thus $w^k = (\lambda^k, \mu^k)$.

Next calculate x^k to maximize the Lagrangean over X. That is, compute

$$\max \left\{ L(x, \lambda^k) = f(x) + \sum_{i=1}^{m} \lambda_i^k g_i(x) \,|\, x \in X \right\} \qquad (14.33)$$

Because X is compact and the f and g_i are continuous, x^k exists.

The solution test is to see if

$$\mu^k = L(x^k, \lambda^k)$$

If w^k does not pass the solution test, then via (14.32) and (14.33)

$$\mu^k < L(x^k, \lambda^k)$$

The calculation of x^k also comprises the map Δ. Specifically

$$\Delta(w^k) = \{x^k \mid x^k \quad \text{maximizes} \quad L(x, \lambda^k) \quad \text{over} \quad X\}$$

By Chap. 7 as X is compact, Δ is closed.
    Finally, we define the set $H(x^k)$ for $x^k \in \Delta(w^k)$ by

$$H(x^k) = \{(\lambda, \mu) \mid f(x^k) + \lambda^t g(x^k) - \mu \leq 0\}$$

    In brief, given z^k, calculate (λ^k, μ^k) to solve subproblem (14.32). Then employing (14.33), we determine x^k. If $\mu^k = L(x^k, \lambda^k)$ stop, otherwise set

$$z^{k+1} = z^k \cap H(x^k)$$

We now prove convergence.
    Condition 1: For $w^k = (\lambda^k, \mu^k)$ it is sufficient to show that the μ^k and the components of λ^k are bounded. First define the point x' by

$$f(x') = \max \{f(x) \mid x \in X\} \tag{14.34}$$

where x' exists because X is compact.
    Certainly

$$\mu^k \leq f(x')$$

Also, setting $\lambda = 0$, we find

$$f(x^0) \leq \mu^k$$

Therefore the μ^k are bounded.
    From (14.29), the definition of z^1, and the fact that $\lambda \geq 0$

$$\lambda_j g_j(x^0) \leq \sum_{i=1}^{m} \lambda_i g_i(x^0) \leq -f(x^0) + \mu \leq -f(x^0) + f(x')$$

Therefore

$$\lambda_j \leq \frac{-f(x^0) + f(x')}{g_j(x^0)} \tag{14.35}$$

and because $0 \leq \lambda_j$, the λ_j are bounded. Consequently, z^1 is compact. Also note that all $x^k \in \Delta(w^k)$ are on the compact set X.

Condition 2: This condition is trivial.

Condition 3: It was previously observed that Δ is closed. Clearly a and b are continuous.

Condition 4: If $w^k = (\lambda^k, \mu^k)$ does not pass the solution test, then

$$\mu^k < L(x^k, \lambda^k) = f(x^k) + (\lambda^k)'g(x^k)$$

But in that case

$$(\lambda^k, \mu^k) \notin H(x^k) = \{(\lambda, \mu)|f(x^k) + \lambda'g(x^k) - \mu \leq 0\}$$

Also the point (λ', μ') where $\lambda' = 0$, and $\mu' = f(x')$ for x' defined in (14.34) satisfies $(\lambda', \mu') \in z^k \cap H(x^k)$.

The procedure converges.

Interpretation of Convergence

We will now prove that if $w^k = (\lambda^k, \mu^k)$ passes the solution test, which is

$$\mu^k = L(x^k, \lambda^k)$$

then we will have determined a solution to problem (14.27).

Let us rewrite the subproblem (14.32).

$$\min \quad \mu$$

$$f(x^i) + \lambda'g(x^i) - \mu \leq 0 \qquad i = 0, \ldots, k-1 \qquad (14.36)$$

$$\lambda \geq 0$$

The dual of the LP problem (14.36) is

$$\max \quad \sum_{i=0}^{k-1} \gamma_i f(x^i)$$

$$\sum_{i=0}^{k-1} \gamma_i g(x^i) \geq 0$$

$$\sum_{i=0}^{k-1} \gamma_i \quad = 1 \qquad (14.37)$$

$$\gamma_i \geq 0$$

where

$$g(x^i) = \begin{pmatrix} g_1(x^i) \\ \cdot \\ \cdot \\ \cdot \\ g_m(x^i) \end{pmatrix}$$

and the γ_i are variables. Via LP duality theory the optimal objective-function value of (14.37) is μ^k.

Examining (14.37), from the concavity of the objective function and constraints, and by property (g) of Chap. 2, the point

$$y^k = \sum_{i=0}^{k-1} \gamma_i x^i \tag{14.38}$$

satisfies

$$f(y^k) \geq \sum_{i=0}^{k-1} \gamma_i f(x^i) \tag{14.39}$$

Selecting the optimal γ_i for (14.37) we find that $f(y^k)$ is at least as great as the optimal objective value of (14.37). Mathematically,

$$f(y^k) \geq \mu^k \tag{14.40}$$

Moreover, y^k is feasible for the primal problem (14.27). This follows because in the same manner as (14.39) was obtained from (14.38)

$$g(y^k) \geq \sum_{i=0}^{k-1} \gamma_i g(x^i) \geq 0 \tag{14.41}$$

Also X is convex so that $y^k \in X$ using property (f) of Chap. 2.

From these observations we obtain an important result. Letting x^* be the optimal solution to problem (14.27), because y^k is feasible, (14.40) ensures

$$f(x^*) \geq \mu^k \tag{14.42}$$

Now the points x^k for $k \geq 1$ are calculated to maximize

$$L(x, \lambda^k) = f(x) + \sum \lambda_i^k g_i(x)$$

over $x \in X$. Therefore, the point (x^k, λ^k) is feasible for the dual, and in fact,

$$L^*(\lambda^k) = L(x^k, \lambda^k) = f(x^k) + \sum_{i=1}^{m} \lambda_i^k g_i(x^k) \qquad (14.43)$$

From Equations (14.26), (14.42), and (14.43) we obtain the result that

$$\mu^k \leq f(x^*) \leq L(x^k, \lambda^k) \qquad (14.44)$$

It follows immediately that if w^k satisfies the solution test,

$$\mu^k = f(x^*) = L(x^k, \lambda^k) \qquad (14.45)$$

Behavior of the Variables

Now examine the points y^k. From (14.39) and (14.40)

$$\mu^k \leq f(y^k) \leq f(x^*)$$

Consequently the limit y^∞ of any convergent subsequence of the y^k,

$$y^k \longrightarrow y^\infty \qquad k \in \mathcal{K}$$

must be an optimal point to problem (14.27) (exercise 14.8).

Duality Interpretation

Using problem (14.36) we now show that the algorithm approximates the dual problem.

The dual problem is

$$\min_{\lambda \geq 0} \left\{ \max_{x \in X} \ L(x, \lambda) \right\}$$

Suppose that $\max_{x \in X} L(x, \lambda)$ were not calculated over all X but only over k points $x^0, \ldots, x^{k-1}$ in X. Then we obtain an approximation to the dual problem

$$\min_{\lambda \geq 0} \left\{ \max_{i=0,\ldots,k-1} L(x^i, \lambda) \right\} \qquad (14.46)$$

Let $\mu = \max_{i=0,\ldots,k-1} L(x^i, \lambda)$; then

$$L(x^i, \lambda) \leq \mu \quad \text{for all} \quad i$$

Problem (14.46) may then be rewritten as

$$\text{min} \quad \mu$$
$$\text{subject to} \quad L(x^i, \lambda) \leq \mu \qquad i = 0, \ldots, k-1 \qquad (14.47)$$
$$\lambda \geq 0$$

However, problem (14.47) is simply problem (14.36).

14.1. For the concave method let h_i, $i = 1, \ldots, n$, be continuously differentiable concave functions. Suppose

$$g(x) = \quad \text{min} \quad [h_1(x), \ldots, h_n(x)]$$

Define

$$u(w) = \nabla h_j(w)$$

where j is any index i such that

$$g(w) = h_i(w)$$

Prove that
(a) $u(w)$ satisfies (14.19)
and
(b) The map

$$\Delta(w) = \{w, u(w)\}$$

is closed at any $w \in U$.
   Hint: Let $w^k \rightarrow w^\infty$, $k \in \mathcal{K}$. There must be a $\mathcal{K}^1 \subset \mathcal{K}$ such that for some j, say $j = 1$,

$$g(w^k) = h_1(w^k) \quad \text{for} \quad k \in \mathcal{K}^1$$

and

$$u(w^k) = \nabla h_1(w^k)$$

Then by continuity

$$g(w^\infty) = h_1(w^\infty)$$

and

$$u(w^\infty) = \nabla h_1(w^\infty)$$

(c) The points $v^k \in \Delta(w^k)$ are on a compact set.

Hint: $v^k = (w^k, u(w^k)) \in \Delta(w^k)$. Consider a sequence $\{v^k\}_{\mathscr{K}}$. By compactness of U for $\mathscr{K}' \subset \mathscr{K}$

$$w^k \longrightarrow w^\infty \qquad k \in \mathscr{K}'$$

Use the hint in (a) above to prove that there is a $\mathscr{K}'' \subset \mathscr{K}'$ such that

$$u(w^k) \longrightarrow u(u^\infty)$$

Then $\{w^k, u(w^k)\}_{\mathscr{K}''}$ converges.

14.2. For the concave method prove that the $u(w)$ are on a compact set and $\Delta(w) = \{w, u(w)\}$ is closed.

Hint: Consider $\{w^k, u(w^k)\}_{\mathscr{K}}$. Because of uniform boundedness, there is a $\mathscr{K}' \subset \mathscr{K}$ such that

$$u(w^k) \longrightarrow u^\infty \qquad k \in \mathscr{K}'$$

Also, because U is compact,

$$w^k \longrightarrow w^\infty \qquad k \in \mathscr{K}'' \subset \mathscr{K}'$$

and by continuity of g

$$g(x) \leq g(w^\infty) + (u^\infty)^i(x - w^\infty)$$

Thus

$$(w^\infty, u^\infty) \in \Delta(w^\infty)$$

14.3. Where in the concave method was the assumption used that the objective function f and constraints g_i be concave? Show by example that this hypothesis is necessary.

14.4. Prove that if $f(x)$ is concave in E^n, then $f(x) - w$ is concave in E^{n+1}.

14.5. Let g_i, $i = 1, \ldots, n$, be continuously differentiable quasi-concave constraints. Suppose u is on the boundary of the feasible region F. Define

$$H(u) = \{x \mid \nabla g_j(u)^i(x - u) \geq 0\}$$

where $\nabla g_j(u) \neq 0$. Then prove that

$$F \subset H(u)$$

14.6. The supporting-hyperplane method required the determination of the point u on the boundary of F such that u was on the line segment between a and w. The point $a \in F$ and $w \notin F$.

Form the function

$$g(x) = \min \ [g_1(x), \ldots, g_m(x)]$$

By assumption $g(a) > 0$, and because $w \notin F$, $g(w) < 0$.

(a) Prove if $g_j(u) = 0$, then assuming $\nabla g_j(u) \neq 0$ ensures that there is a unique boundary point u such that

$$g_j(u) = 0$$

(b) Modify the golden-section and bisecting-search procedures to determine u.

14.7. Prove condition 4 for the supporting-hyperplane method.

   Hint: $g_j(w) < 0$, $g_j(a) > 0$, $g_j(u) = 0$, and $\nabla g_j(u) \neq 0$. Reduce the problem to E^1 by considering the line between a and w.

14.8. Prove in the dual-cutting-plane algorithm that if

$$y^k \longrightarrow y^\infty \qquad k \in \mathcal{K}$$

where y^k is defined in (14.38), then y^∞ is an optimal point for problem (14.27).

14.9. In the dual-cutting-plane algorithm show how to determine at each iteration k a good upper bound for the $f(x^*)$, the optimal value of problem (14.27).

14.10. Consider the dual method. In computer practice the set X is taken to be E^n. Then x^k is calculated to

$$\max_{x \in E^n} \ L(x, \lambda^k)$$

Suppose the f and g_i were separable; that is, let $x = (x_1, x_2)$ where $x_1 \in E^{n_1}$, $x_2 \in E^{n_2}$, and $n_1 + n_2 = n$. Then

$$f(x) = f^1(x_1) + f^2(x_2)$$

and

$$g_i(x) = g_i^1(x_1) + g_i^2(x_2)$$

Indicate how the above problem of calculating x^k simplifies.

14.11. For the dual-cutting-plane method, prove

$$(a) \quad \mu^k \leq \mu^{k+1}$$

and

$$(b) \quad \lim_{k \to \infty} \mu^k = f(x^*)$$

NOTES AND REFERENCES

§14.1. The development of cutting-plane methods from this viewpoint is new.

§14.2. The algorithm was suggested by KELLEY and by CHENEY and GOLD-STEIN see also WOLFE (1961). The proof via the Cutting-Plane Convergence Theorem is new.

§14.3. This algorithm is based upon the article by VEINOTT (1967), although the proof is new.

§14.4. The algorithm in this section is the dual of a generalized programming algorithm developed by Dantzig and Wolfe [see DANTZIG (1963), Chap. 24]. The original presentation was fundamentally different, being based upon column generation and not cutting planes.

Appendix

Mathematical and Linear-

Programming Review

This appendix summarizes several mathematical and linear programming (LP) topics. A more detailed review can be obtained from APOSTLE; FLEMING; FINKBEINER; HALMOS; and HADLEY (1961) for the mathematical topics and from CHARNES and COOPER (1961); DANTZIG (1963); GALE; GASS; HADLEY (1962); RILEY and GASS; SIMONNARD; and FARKAS for the LP topics.

A.1. MATHEMATICAL TOPICS

Column Vectors

All vectors in the text are column vectors. Thus if $x \in E^n$, where E^n indicates n-dimensional Euclidean space,

$$x = \begin{pmatrix} x_1 \\ \vdots \\ x_n \end{pmatrix} \quad \text{or} \quad x = (x_1, \ldots, x_n)^t$$

The statement $x \geq 0$ means $x_i \geq 0$ for all i; while $x \neq 0$ means $x_i \neq 0$ for at least one i. The superscript t denotes transpose.

The notation "$\in$" indicates is an element of; similarly, "$\notin$" indicates is not an element of.

Distance and Limits

Consider now points in E^n. The Euclidean distance between $x = (x_i)$ and $y = (y_i)$ is

$$\| x - y \| = \left(\sum_{i=1}^{n} (x_i - y_i)^2 \right)^{1/2}$$

Given a sequence of points $\{x^k\}_1^\infty$, $x^k \in E^n$, we say the sequence converges to a limit $x^\infty \in E^n$, denoted

$$\lim_{k \to \infty} x^k = x^\infty$$

or

$$x^k \longrightarrow x^\infty \qquad k \longrightarrow \infty$$

if the following holds. Given $\epsilon > 0$, there is a P such that $k > P$ implies
that

$$\| x^k - x^\infty \| < \epsilon$$

Essentially, the distance between x^k and x^∞ gets arbitrarily small as k gets
large.

Several properties of limits are listed. Suppose $x^k \longrightarrow x^\infty$ and $y^k \longrightarrow y^\infty$,
where all points are in E^n. Then

$$\lim_{k \to \infty} (x^k + y^k) = x^\infty + y^\infty$$

and

$$\lim_{k \to \infty} [(x^k)^t y^k] = (x^\infty)^t y^\infty$$

If all points are in E^1,

$$\lim_{k \to \infty} \frac{x^k}{y^k} = \frac{x^\infty}{y^\infty} \quad \text{if} \quad y^\infty \neq 0$$

A Function

Let a function $h: E^n \longrightarrow E^m$ indicate that h takes points $x \in E^n$ into
points $h(x)$ in E^m. The function $h(x)$ is itself a vector $h(x) = (h(x)_1, \ldots, h(x)_m)^t$. In other words, h is a vector-valued function.

Continuity

A function $h: E^n \longrightarrow E^m$ is continuous at x^∞ if

$$\lim_{k \to \infty} x^k \longrightarrow x^\infty$$

implies

$$\lim_{k \to \infty} h(x^k) = h(x^\infty)$$

An equivalent definition is: Given $\epsilon > 0$ there exists a $\delta > 0$ such that

$$\| x - x^\infty \|_n < \delta$$

implies

$$\| h(x) - h(x^\infty) \|_m < \epsilon$$

Here

$$\|x\|_n = \left(\sum_{i=1}^{n} (x_i^2)\right)^{1/2} \quad \text{and} \quad \|h(x)\|_m = \left(\sum_{i=1}^{m} (h(x)_i)^2\right)^{1/2}$$

A function is continuous if it is continuous at all points x for which it is defined.

Composition of Continuous Functions

Let $h: E^{m_1} \rightarrow E^{m_2}$ and $l: E^{m_2} \rightarrow E^{m_3}$ be continuous functions. Then the composite function $j: E^{m_1} \rightarrow E^{m_3}$ defined by

$$j(x) = l[h(x)]$$

is also continuous.

Partial Derivatives

Let $h: E^n \rightarrow E^1$ be given. If h has first partial derivatives, it has a **gradient**

$$\nabla h(x) = \left(\frac{\partial h(x)}{\partial x_1}, \cdots, \frac{\partial h(x)}{\partial x_n}\right)^t$$

Should h have second partial derivatives, it possesses an $n \times n$ Hessian matrix

$$H(x) = \left(\frac{\partial^2 h(x)}{\partial x_i \partial x_j}\right)$$

Example: If $h(x) = q^t x + \frac{1}{2} x^t Q x$,

$$\nabla h(x) = q + Q x$$

and

$$H(x) = Q$$

Taylor's Expansion

If $h: E^n \rightarrow E^1$ has continuous first partial derivatives, it also has a first-order Taylor expansion about any point x^1

$$h(x) = h(x^1) + \nabla f(x^1 + \theta(x - x^1))^t (x - x^1)$$

where $0 \leq \theta \leq 1$. The point $x^1 + \theta(x - x^1)$ is some point on the line between x^1 and x.

Should h have continuous second partial derivatives, it possesses a second-order Taylor expansion

$$h(x) = h(x^1) + \nabla f(x^1)'(x - x^1) + \tfrac{1}{2}(x - x^1)'H(x^1 + \theta(x - x^1))(x - x^1)$$

where $0 \leq \theta \leq 1$.

Directional Derivative

Any vector $d \in E^n$ indicates a direction. Suppose $\|d\| = 1$ so that the direction d is of unit length and $h: E^n \to E^1$ is differentiable. Then the directional derivative at x in the direction d is

$$\lim_{\tau \to 0} \frac{h(x + \tau d) - h(x)}{\tau} = \nabla h(x)'d$$

The directional derivative is the instantaneous rate of change in h at x in the direction d.

Chain Rule

Let $h: E^{m_1} \to E^{m_2}$ and $l: E^{m_2} \to E^1$ have continuous first partial derivatives. Then the composite function $j: E^{m_1} \to E^1$, where

$$j(x) = l[h(x)]$$

has partial derivatives

$$\frac{\partial j(x)}{\partial x_i} = \sum_{k=1}^{m_2} \frac{\partial l}{\partial y_k} \frac{\partial h_k(x)}{\partial x_i} \qquad i = 1, \ldots, m_1$$

in which $\partial l/\partial y_k$ means the kth partial derivative of l evaluated at $h(x)$.

Continuity Implied by Differentiability

If $h: E^n \to E^1$ has continuous first partial derivatives, it is also continuous. Should h have continuous second partial derivatives, it, in addition, possesses continuous first partial derivatives. Consequently, continuous second partial derivatives imply both continuity and continuous first partial derivatives.

Closed Sets

Let X be a set in a space V. Then X is closed if, given a sequence $x^k \in X$ such that

$$x^k \longrightarrow x^\infty \qquad k \longrightarrow \infty$$

then

$$x^\infty \in X$$

Thus a set is closed if, given a convergent sequence of points from the set, the limit of the sequence is itself in the set.

Given a set X that may not be closed, we may form its closure $\bar{X}$. The closure $\bar{X}$ consists of all limits of convergent sequences from X. The closure set $\bar{X}$ is a closed set. If X is itself closed,

$$X = \bar{X}$$

Examples

$$\{x \mid 0 \leq x \leq 1\} \subset E^1 \quad \text{is closed}$$
$$E^1 \quad \text{is closed.}$$

The set $X = \{x \mid 0 < x\}$ is not closed. Its closure is

$$\bar{X} = \{x \mid 0 \leq x\}$$

Hyperplane and Half Space

A hyperplane in E^n is a set of the form

$$\mathscr{H} = \{x \mid c^t x = \alpha \quad \text{where } c \in E^n,\ c \neq 0,\ \alpha \in E^1\}$$

The hyperplane determines two (closed) half spaces

$$\mathscr{H}^+ = \{x \mid c^t x \geq \alpha\}$$

and

$$\mathscr{H}^- = \{x \mid c^t x \leq \alpha\}$$

A.2. LINEAR PROGRAMMING

The programming problem

$$\text{subject to} \quad \max \quad q'x$$
$$Ax = b \tag{A.1}$$
$$x \geq 0$$

where A is an $m \times n$ matrix is a LP problem.

The Dual Theorem of Linear Programming

The dual problem is to

$$\text{subject to} \quad \min \quad b'u$$
$$A'u \geq q \tag{A.2}$$
$$u \quad \text{unconstrained}$$

If either problem (A.1) or (A.2) has a finite optimal solution, then so does the other, and the objective-function values are equal. The two problems are termed dual problems.

A different pair of dual problems is

$$\max \quad q'x$$
$$Ax \leq b \tag{A.3}$$

and

$$\min \quad b'u$$
$$A'u = q \tag{A.4}$$
$$u \geq 0$$

Again if either one has a finite optimal solution, so does the other, and the respective objective functions are equal. The two pairs of dual problems are equivalent.

Farkas' Lemma

Farkas' Lemma is the following theorem.

Theorem: *The statement*

$$q^t x \leq 0 \tag{A.5}$$

$$\textit{for all} \quad x \quad \textit{such that} \quad Ax \geq 0$$

is equivalent to the statement that there exists $u \geq 0$ such that

$$q + A^t u = 0 \tag{A.6}$$

PROOF.   Consider the problem

$$\begin{aligned} \max \quad & q^t x \\ & Ax \geq 0 \end{aligned} \tag{A.7}$$

The dual is

$$\begin{aligned} \min \quad & 0 \\ & -A^t u = q \\ & u \geq 0 \end{aligned} \tag{A.8}$$

If statement (A.5) holds, then, letting $x = 0$, we find problem (A.7) has a finite optimal solution with objective-function value of zero. Duality provides that problem (A.8) has a solution, but if (A.8) has a solution, statement (A.6) holds.

Conversely, suppose statement (A.6) holds. Then problem (A.8) has a finite optimal solution since the u satisfying (A.6) gives a value of zero to the objective function of (A.8). Via duality theory (A.7) has an optimal solution with objective function of zero. But if (A.7) has an optimal objective-function value of zero, then (A.5) holds. ◆

The Simplex Method

The simplex method for solving the LP problem will now be briefly described. The LP problem is

$$\begin{aligned} \max \quad & q^t x \\ & Ax = b \\ & x \geq 0 \end{aligned}$$

where A is $m \times n$.

The simplex method operates on certain matrices T termed tableaux. Each tableau T has the property that

$$T = C^{-1} A$$

where C is an $m \times m$ basis matrix composed of m linearly independent columns of A. Thus if $A = (a^1, \ldots, a^m)$, letting a^i indicate a column of A,

$$C = [a^{B_1}, a^{B_2}, \ldots, a^{B_m}]$$

where B_j indicates a column number from A. The columns of A comprising C form the basis and are said to be basic. Observe that the B_jth column of T is a column of zeroes except for a one in the jth place.

Letting $b' == C^{-1}b$ we see that any solution to

$$Tx = b'$$

is a solution to

$$Ax = b$$

and conversely.

A basic feasible solution is a point x such that

$$Tx = b'$$
$$x \geq 0$$

where in addition

$$x_{B_j} = b'_j \qquad j = 1, \ldots, m$$

and all other x components are zero. The x_{B_j} correspond to the columns of A in the basis, and these x components are also said to be basic. Note that $b'_i \geq 0$ for all i.

Clearly, a basic feasible solution corresponds to a given tableau T.

Simplex-Method Operation

To specify the operation of the simplex method let

$$q_B = (q_{B_1}, \ldots, q_{B_m})^t$$

Suppose a basic feasible point x^1 with corresponding tableau is given. Calculate the relative-cost vector $c = (c_i)$ where

$$c = (q - T^t q_B)$$

and let $c_s = \max \{c_i \mid i = 1, \ldots, m\}$. If $c_s \leq 0$, then x^1 is optimal.

Otherwise increase x_s adjusting only basic variables until a basic variable, say x_{B_r}, becomes zero. The corresponding x value is x^2. Pivot as discussed below. Replace x_{B_r} with x_s in the basis and continue from x^2 as before. Also a^s replaces a^{B_r} in the basis so that the new $B_r = s$.

The variable x^2 is also a basic feasible solution. Its corresponding tableau can be formed by pivoting on the element in the old tableau that is in row r, column s. The pivot procedure amounts to adding multiples of row r in the old tableau until there is a one in the rth position of column s and zeroes elsewhere in that column.

By pivoting in this manner we obtain the tableau corresponding to the new basis matrix C. The new basis matrix is the same as the old one except that now the rth column of C contains a^s.

To determine the element r of row s on which to pivot, perform the calculation

$$\frac{b_r^1}{t_{rs}} = \min \left\{ \frac{b_i^1}{t_{is}} \,\middle|\, t_{is} > 0 \right\}$$

where t_{is} is the element in the ith row of column s, and b^1 is the right-hand side corresponding to the tableau.

If no $t_{is} > 0$, then the LP problem has an unbounded solution. In effect, x_s can be increased indefinitely always increasing the objective-function value. Geometrically, increasing x_s and adjusting only basic variables forms a ray starting from x^1 and extending indefinitely.

Bibliography

Abadie, J. (ed.) (1967), *Nonlinear Programming*. John Wiley & Sons, Inc., New York.

Apostol, T. M. (1957), *Mathematical Analysis*. Addison-Wesley Publishing Co., Inc., Reading, Mass.

Arrow, K. J. (1951), "A Gradient Method for Approximating Saddle Points and Constrained Maxima," Paper P-223, The RAND Corporation, Santa Monica, Calif.

Arrow, K. J., and A. C. Enthoven (1961), "Quasi-Concave Programming," *Econometrica*, XXIX, No. 4, 779–800.

Arrow, K. J., and L. Hurwicz (1956), "Reduction of Constrained Maxima to Saddle Point Problems," *Proceedings of the Third Berkeley Symposium on Mathematical Statistics and Probability*, ed. J. Neyman, University of California Press, Berkeley, pp. 1–20.

———— (1960), "Decentralization and Computation in Research Allocation," pp. 34–104 in *Essays in Economics and Econometrics*. University of North Carolina Press, Chapel Hill.

Arrow, K. J., L. Hurwicz, and H. Uzawa (1958), *Studies in Linear and Non-Linear Programming*. Stanford University Press, Stanford, Calif.

———— (1961), "Constraint Qualification in Maximization Problems," *Naval Research Logistics Quarterly*, VIII, 175–91.

Arrow, K. J., and H. Uzawa (1960), "Constraint Qualification in Maximization Problems, II," Technical Report No. 84, Institute for Mathematical Studies in Social Sciences, Stanford University, Stanford, Calif.

Avriel, M., and D. J. Wilde (1966), "Optimal Condenser Design by Geometric Programming," *Stanford Chemical Engineering Report.*

Barankin, E. W., and R. Dorfman (1955), "On Quadratic Programming." *University of California Publication in Statistics*, II, 285–318, University of California, Berkeley.

Baumol, W. J. (1961), *Economic Theory and Operations Analysis.* Prentice-Hall, Inc., Englewood Cliffs, N.J.

Beale, E. M. L. (1955), "On Minimizing a Convex Function Subject to Linear Inequalities," *Journal of the Royal Statistical Society (B)*, XVII, 173–84.

―――― (1959), "An Algorithm for Solving the Transportation Problems when the Shipping Cost over each Route is Convex," *Naval Research Logistics Quarterly*, VI, 43–56.

―――― (1959), "On Quadratic Programming," *Naval Research Logistics Quarterly*, VI, 227–43.

Beckman, F. S. (1960), "The Solution of Linear Equations by the Conjugate Gradient Method," pp. 62–72 in *Mathematical Methods for Digital Computers*, eds. A. Ralston and H. S. Wilf. John Wiley & Sons, Inc., New York.

Bellman, R. (1957), *Dynamic Programming*. Princeton University Press, Princeton, N. J.

——— (1961), *Adaptive Control Process: A Guided Tour*. Princeton University Press, Princeton, N. J.

Bellman, R., and S. Dreyfus (1962), *Applied Dynamic Programming*. Princeton University Press, Princeton, N.J.

Beltrami, E. J., and R. McGill (1966), "A Class of Variational Problems in Search Theory and the Maximum Principle," *Operations Research*, XIV, No. 2, 267–78.

Berge, C. (1963), *Topological Spaces*. Trans. by E.M. Patterson. Oliver and Boyd, LTD., Edinburgh and London.

Berge, C., and A. Ghouila-Houri (1965), *Programming, Games and Transportation Networks*. Trans. M. Merrington & C. Ramanujacharyulu. John Wiley & Sons, Inc., New York.

Birkhoff, G., and S. MacLane (1941), *A Survey of Modern Algebra*. The Macmillan Company, Publishers, New York.

Blackwell, D. (1962), "Discrete Dynamic Programming," *Annuals of Mathematical Statistics*, XXXIII, 719–26.

Boot, J. C. G. (1963), "On Sensitivity Analysis in Convex Quadratic Programming Problems," *Operations Research*, XI, No. 5, 771–86.

——— (1964), *Quadratic Programming*. Rand McNally & Company, Chicago.

Bregman, L. M. (1963), "The Method of Successive Projection for Finding a Common Point of Convex Sets," *Soviet Mathematics*, VI, 688–92.

Brinkley, S. R. (1947), "Calculation of the Equilibrium Composition of Systems of Many Constituents," *Journal of Chemical Physics*, XV, 107–10.

Brooks, S. H. (1959), "A Comparison of Maximum-Seeking Methods," *Operations Research*, VII, No. 4, 430–57.

Bui-Trong-Lieu and P. Huard (1966), "La Méthode des Centres dans un Espace Topologique," *Numerische Mathmatik*, III, No. 1, 56–67.

Burger, E. (1955), "On Extrema with Side Conditions," *Econometrica*, XXIII, No. 4, 451–52.

Butler, T., and A. V. Martin (1962), "On a Method of Courant for Minimizing Functionals," *Journal of Mathematics and Physics*, XLI, 291–99.

Camp, G. D. (1955), "Inequality-Constrained Stationary Value Problems," *Journal of the Operations Research Society of America*, III, 548–50.

Canon, M. D., and C. D. Cullum (1967), "A Tight Upper Bound on the Rate of Convergence of the Frank-Wolfe Algorithm," IBM Watson Research Center, Yorktown Heights, N. Y. (Mimeo.).

Canon, M. D., C. D. Cullum, and E. Polak (in press), *Optimization, Control, and Algorithms*. McGraw-Hill Book Company, Publishers, New York.

Carroll, C. W. (1961), "The Created Response Surface Technique for Optimizing Non-Linear Restrained Systems," *Operations Research*, IX, No. 2, 169–84.

——— (1959), *An Operations Research Approach to the Economic Optimization of a Kraft Pulping Process*. The Institute of Paper Chemistry, Appleton, Wis.

Cauchy, A. L. (1847), "Méthode Générale pour la Résolution des Systèmes d'Équations Simultanées," *Comptes Rendus, Académie Science, Paris*, XXV, 536–38.

Cesari, L. (1964), *Problemi di Lagrange con Vincoli Unilaterali*. Academia delle Science di Torino Atti 98, Supplement, 88–119.

Charnes, A., and W. W. Cooper (1959), "Chance-Constrained Programming," *Management Science*, VI, 73–9.

Charnes, A., and W. W. Cooper (1961), *Management Models and Industrial Applications of Linear Programming*. 2 vols. John Wiley & Sons, Inc., New York.

Charnes, A., W. W. Cooper, and K. Kortanek (1962), "Duality, Haar Programms and Finite Sequence Spaces," *Procedings of the National Academy of Science*, XLVIII, 783.

Charnes, A., and C. Lemke (1954), "Minimization of Nonlinear Separable Convex Functionals," *Naval Research Logistics Quarterly*, I, 301–12.

Cheney, E. W., and A. A. Goldstein (1959), "Newton's Method of Convex Programming and Tchebycheff Approximation," *Numerische Mathmatik*, I, 253–68.

Chevassus, M. (1967), "Condition Suffisante de Convergence pour des Methods Itératives de Minimisation," No. HR 7713, *Electricité de France*.

Collatz, L. (1966), *Functional Analysis and Numerical Mathematics*, Chap. 27. Academic Press Inc., New York.

Connors, M. M., and D. Teichroew (1967), *Optimal Control of Dynamic Operations Research Models*. International Textbook Co., Scranton, Pa.

Cottle, R. W. (1963a), "A Theorem of Fritz John in Mathematical Programming," Research Memorandum RM-3858-PR, The RAND Corporation, Santa Monica, Calif.

——— (1963b), "Symmetric Dual Quadratic Programs," *Quarterly of Applied Mathematics*, XXI, 237–43.

Cottle, R. W., and G. B. Dantzig (1967), "Complementary Pivot Theories of Mathematical Programming," Technical Report No. 16, Department of Operations Research, Stanford University, Stanford, Calif.

Courant, R. (1943), "Variational Methods for the Solution of Problems of Equilibrium and Vibrations," *Bulletin American Mathematics Society*, XLIX, 1–23.

Coxeter, H. S. M. (1954), "The Golden section, Phyllotaxis, and Wythoff's game," *Scripta Mathematica*, 135–43.

Crockett, J. B., and H. Chernoff (1955), "Gradient Methods of Maximization," *Pacific Journal of Mathematics*, V, 33–50.

Curry, H. (1944), "The Method of Steepest Descent for Non-linear Minimization Problems," *Quarterly of Applied Mathematics*, II, 258–61.

Dantzig, G. B. (1955), "Linear Programming Under Uncertainty," *Management Science*, I, 197–206.

——— (1963), *Linear Programming and Extensions*. Princeton University Press, Princeton, N.J.

Dantzig, G. B., E. Eisenberg, and R. W. Cottle (1965), "Symmetric Dual Nonlinear Programs," *Pacific Journal of Mathematics*, XV, 809–12.

Dantzig, G. B., S. Johnson, and W. White (1958), "A Linear Programming Approach to the Chemical Equilibrium Problem," *Management Science*, V, 38–43.

Davidon, W. C. (1959), "Variable Metric Method for Minimization," A.E.C. Research and Development Report, ANL-5990 (Rev.).

Debreu, G. (1954), "Representation of a Preference Ordering by a Numerical Function," pp. 159–65 in eds. R. M. Thrall, C. H. Cooms, and R. L. Davis, *Decision Processes*. John Wiley & Sons, Inc., New York.

——— (1967), "Integration of Correspondences," *Proceedings of the Fifth Berkeley Symposium on Mathematical Statistics and Probability*, II, Part 1, eds. L. Le Com and J. Neyman, University of California Press, Berkeley, 351–72.

Dennis, J. B. (1959), *Mathematical Programming and Electrical Networks*. MIT Press, Cambridge, Mass.

d'Epenoux, F. (1960), "Sur un problème de production et de stockage dans l'aléatoire," *Revue Française de Recherche Opérationnelle*, IV, No. 1, 3–15. ["A Probabilistic Production and Inventory Problem," *Management Science*, X (1963) 98–108. (A translation with some modifications of the French original.)]

D'Esopo, D. (1959), "A Convex Programming Procedure," *Naval Research Logistics Quarterly*, VI, No. 1, 33–42.

Dorn, W. S. (1960a), "A Duality Theorem for Convex Programs," *IBM Journal Research & Development*, IV, 407–13.

——— (1960b), "A Symmetric Dual Theorem for Quadratic Programs," *Journal of Operations Research Society of Japan*, II, No. 93.

———. (1960c), "Duality in Quadratic Programming," *Quarterly of Applied Mathematics*, XVIII, 155–62.

——— (1961), "Self-Dual Quadratic Programs," *Journal of the Society for Industrial and Applied Mathematics*, IX, 51–4.

——— (1963), "Non-Linear Programming—A Survey," *Management Science*, IX, 171–208.

Duffin, R. J., E. L. Peterson, and C. M. Zener (1967), *Geometric Programming*. John Wiley & Sons, Inc., New York.

Eggleston, H. G. (1963), *Convexity*. Cambridge Tracts in Mathematics and Mathematical Physics, No. 47, Cambridge University Press, London.

Eremin, I. I. (1965), "The Relaxation Method of Solving Systems of Inequalities with Convex Functions on the Left Sides," *Soviet Mathematics*, VI, 219–22.

Farkas, J. (1901), "Über die Theorie der einfachen Ungleichungen," *Journal für die reine und angewandte Mathematik*, CXXIV, 1–27.

Fenchel, W. (1953), "Convex Cones, Sets, and Functions," Lecture Notes, Department of Mathematics, Princeton University.

Fiacco, A. V. (1967), "Sequential Unconstrained Minimization Methods for Nonlinear Programming." Unpublished Ph. D. dissertation, Northwestern University, Evanston, Ill.

Fiacco, A. V., and G. P. McCormick (1963), "Programming Under Nonlinear Constraints by Unconstrained Minimization: A Primal-Dual Method," RAC-TP-96, Research Analysis Corp., McLean, Va.

—— (1964a), "Computational Algorithm for the Sequential Unconstrained Minimization Technique for Nonlinear Programming," *Management Science,* X, No. 2, 601–17.

—— (1964b), "The Sequential Unconstrained Minimization Technique for Nonlinear Programming, A Primal-Dual Method," *Management Science,* X, No. 2, 360–66.

—— (1965a), "SUMT Without Parameters," System Research Memo. No. 121, Tech. Inst., Northwestern University, Evanston, Ill.

—— (1965b), "The Sequential Unconstrained Minimization Technique for Convex Programming with Equality Constraints," RAC-TP-155, Research Analysis Corp., McLean, Va.

—— (1966), "Extensions of SUMT for Nonlinear Programming Equality Constraints and Extrapolation," *Management Science,* XII, No. 11, 816–28.

Finkbeiner, II, D. T. (1960), *Introduction to Matrices and Linear Transformations.* W.H. Freeman and Co., Publishers, San Francisco.

Fleming, W. (1965), *Functions of Several Variables.* Addison-Wesley Publishing Co., Inc., Reading, Mass.

Fletcher, R. (1965), "Function Minimization Without Evaluating Derivatives: A Review," *The Computer Journal,* VIII, No. 1, 33–41.

Fletcher, R., and M. J. D. Powell (1963), "A Rapidly Convergent Descent Method for Minimization," *The Computer Journal,* VI, 163.

Fletcher, R., and C.M. Reeves (1964), "Function Minimization by Conjugate Gradients," *The Computer Journal,* VII, No. 2, 149–54.

Forsythe, G. E. (1955), "Computing Constrained Minima with Lagrange Multipliers," *Journal of the Society for Industrial and Applied Mathematics,* III, 173–78.

Forsythe, G. E. (1967), "On the Asymptotic Directions of the s-Dimensional Optimum Gradient Method," Technical Report No. CS 61, Computer Science Department, Stanford University, Stanford, Calif.

Frank, M., and P. Wolfe (1956), "An Algorithm for Quadratic Programming," *Naval Research Logistics Quarterly,* III, 95–110.

Frisch, K. R. (1955), "The Logarithmic Potential Method of Convex Programming," (Memo.) Univ. Inst. of Economics, Oslo.

Fuare, P., and P. Huard (1965), "Résolution de Programmes Mathématiques à Fonction Non Linéaire par la Méthode du Gradient Réduit," *Revue Française de Recherche Opérationnelle,* IX, No. 36, 167–206.

Gale, D. (1960), *The Theory of Linear Economic Models.* McGraw-Hill Book Company, New York.

Gass, S. I. (1958), *Linear Programming, Methods and Applications.* McGraw-Hill Book Company, New York.

Geoffrion, A. M. (1966), "Strictly Concave Parametric Programming, Part I: Basic Theory," *Management Science,* XIII, No. 3, 244–53.

—— (1967), "Strictly Concave Parametric Programming, Part II: Additional Theory and Computational Considerations," *Management Science,* XIII, No. 5, 359–70.

Gibbs, J. W. (1961), "On the Equilibrium of Heterogeneous Substances," *The Scientific Papers of J. Willard Gibbs.* I, Dover Publications, Inc., New York.

Glover, F. (1966), "Truncated Enumeration Methods for Solving Pure and Mixed Integer Linear Programs," Working Paper No. 27, Operations Research Center, University of California, Berkeley.

Goldstein, A. (1962), "Cauchy's Method of Minimization," *Numerische Mathematik,* IV, 146–50.

——— (1965a), "Convex Programming and Optimal Controls," *SIAM Journal on Control,* III, No. 1, 142–46.

——— (1965b), "On Steepest Descent," *SIAM Journal on Control,* III, No. 1, 147–51.

——— (1966), "Minimizing Functionals on Normed Linear Spaces," *SIAM Journal on Control,* IV, No. 1, 194–210.

Goldstein, A., and B. R. Kripke (1964), "Mathematical Programming by Minimizing Differentiable Functions," *Numerische Mathematik,* VI, 47–8.

Graves, R. L., and P. Wolfe (1963), *Recent Advances in Mathematical Programming,* McGraw-Hill Book Company, New York.

Hadley, G. F. (1960), "How Practical is Nonlinear Programming?" *Product Engineering,* XXXI, 78–80.

——— (1961), *Linear Algebra.* Addison-Wesley Publishing Co., Inc., Reading, Mass.

——— (1962), *Linear Programming.* Addison-Wesley Publishing Co., Inc., Reading, Mass.

——— (1964), *Nonlinear and Dynamic Programming.* Addison-Wesley Publishing Co., Inc., Reading, Mass.

Halmos, P. R. (1958), *Finite Dimensional Vector Spaces.* (2nd ed.) D. Van Nostrand Company, Inc., Princeton, N.J.

Hancock, H. (1960), *Theory of Maxima and Minima.* Dover Publications, Inc., New York.

Hanson, M. A. (1961), "A Duality Theorem for Nonlinear Programming with Nonlinear Constraints," *Australia Journal of Statistics,* III.

Hestenes, M. R. (1966), *Calculus of Variations and Optimal Control Theory.* John Wiley & Sons, Inc., New York.

Hestenes, M. R., and E. Stiefel (1952), "Methods of Conjugate Gradients for Solving Linear Systems," *Journal of Research of the National Bureau of Standards,* XLIX, 409.

Hildreth, C. (1957), "A Quadratic Programming Procedure," *Naval Research Logistics Quarterly,* XIV, 79–85.

Hillier, F. S., and G. J. Lieberman (1967), *Introduction to Operations Research.* Holden-Day, Inc., Publisher, San Francisco.

Himsworth, F. R., W. Spendley, and G. R. Hext (1962), "The Sequential Application of Simplex Designs in Optimization and Evolutionary Operation," *Technometrics,* IV, 441.

Hitch, C. J., and R. N. McKean (1960), *The Economics of Defense in the Nuclear Age.* Harvard University Press, Cambridge, Mass.

Houthakker, H. (1960), "The Capacity Method of Quadratic Programming," *Econometrica*, XXVIII, No. 1, 62–87.

Howard, R. (1960), *Dynamic Programming and Markov Processes*. MIT Press, Cambridge, Mass.

Hu, T. C. (1964), "Minimum Convex-Cost Flows in Networks," RM-4265-PR, The RAND Corporation, Santa Monica, Calif.

Huard, P. (1962), "Dual Programs," *Journal Research & Development*, VI, 137–39.

–––––– (1967), "Method of Centers," Chap. 7 in *Non-Linear Programming*. ed. J. Abadie, John Wiley & Sons, Inc., New York.

Hurwicz, L. (1960), "Conditions for Economic Efficiency of Centralized and Decentralized Structure," p. 169 in *Value and Plan*. ed. Gregory Grossman, University of California Press, Berkeley and Los Angeles.

Ivanov, V. (1962a), "A General Approximation Method for Solving Linear Problems," *Soviet Mathematics*, III, No. 2, 415–18.

–––––– (1962b), "Algorithms of Rapid Descent," *Soviet Mathematics*, III, No. 2, 476–79.

John, F. (1948), "Extremum Problems with Inequalities as Side Conditions," pp. 187–204 in *Studies and Essays, Courant Anniversary Volume*, ed. K. O. Friedrichs, O. E. Neugebauer, and J. J. Stoker, John Wiley & Sons, Inc. New York.

Kalman, R. E., and J. E. Bertram (1960), "Control System Analysis and Design Via the 'Second Method' of Liapunov, I and II," *Journal of Basic Engineering*, LXXXII, 371–400.

Kantorovich, L. V., and G. P. Akilov (1964), *Functional Analysis in Normed Spaces*, Chap 15. The Macmillan Company, Publishers, New York.

Karamardian, S. (1966), "Duality in Mathematical Programming," ORC 66–2, Operations Research Center, University of California, Berkeley.

Karlin, S. (1959), *Mathematical Methods and Theory in Games, Programming, and Economics*. vols. I and II, Addison-Wesley Publishing Co., Inc., Reading, Mass.

Kelley, J. L. (1963), *General Topology*. D. Van Nostrand Company, Inc., Princeton, N.J.

Kelly, J. E. (1960), "The Cutting-Plane Method for Solving Convex Programs," *Journal of the Society for Industrial and Applied Mathematics*, VIII, No. 4, 703–12.

Kiefer, J. (1953), "Sequential Minimax Search for a Maximum," *Proceedings of the American Mathematics Society*, IV, 502–6.

Koopmans, T. (ed.) (1951), *Activity Analysis of Production and Allocation*. John Wiley & Sons, Inc., New York.

Kuhn, H. W., and A. W. Tucker (1951), "Nonlinear Programming," in *Proceedings of the Second Berkeley Symposium on Mathematical Statistics and Probability*, ed. J. Neyman. University of California Press, Berkeley and Los Angeles, California, 481–92.

Kunzi, H. P., and W. Krelle (1966), *Nonlinear Programming*. Blaisdell Publishing Co., Inc., Waltham, Mass.

LaSalle, J., and S. Lefschetz (1961), *Stability by Liapunov's Direct Method*. Academic Press Inc., New York.

Leitmann, G. (ed.) (1962), *Optimization Techniques: With Applications to Aerospace Systems.* Academic Press Inc., New York.

Lemke, C. E. (1962), "A Method for Solution of Quadratic Programs," *Management Science*, VIII, No. 4, 442–53.

——— (1965), "Bimatrix Equilibrium Points and Mathematical Programming," *Management Science*, XI, No. 11, 681–89.

Lemke, C. E., and J. T. Howson, Jr. (1964), "Equilibrium Points of Bimatrix Games," *Journal of the Society for Industrial and Applied Mathematics*, XII, No. 2, 413–23.

Lhermitte, P., and F. Bessière (1963), "Sur les Possibilités de la Programmation Non Linéaire Appliquée au Choix des Investissements," *Actes de la 3e Conférence Internationale de Recherche Opérationnelle*, Oslo, (Dunod), pp. 597–609.

Liapunov, A. M. (1907), "Problème Général de la Stabilité du Movement," *Ann. Fac. Sci. Toulouse*, IX, 203–475.

Malinvaud, E. (1966), *Statistical Methods of Econometrics.* pp. 310–14, Rand McNally & Company, Chicago, Ill.

Mangasarian, O. L. (1962), "Duality in Nonlinear Programming," *Quarterly of Applied Mathematics*, XX, 300–302.

——— (1965), "Pseudo-Convex Functions," *SIAM Journal on Control*, III, No. 2, 281–90.

Mangasarian, O. L., and S. Fromovitz (1967), "The Fritz John Necessary Optimality Conditions in the Presence of Equality and Inequality Constraints," *Journal of Mathematical Analysis and Application*, XVII, No. 1, 37–47.

Mangasarian, O. L., and J. Ponstein (1965), "Minimax and Duality in Nonlinear Programming," *Journal of Mathematical Analysis and Applications*, XI, No. 1–3, 504–18.

Manne, A. S. (1956), *Scheduling of Petroleum Refinery Operations.* Harvard University Press, Cambridge, Mass.

——— (1960), "Linear Programming and Sequential Decisions," *Management Science*, VI, No. 3, 259–67.

Markowitz, H. (1956), "The Optimization of a Quadratic Function Subject to Linear Constraints," *Naval Research Logistics Quarterly*, III, 111–33.

——— (1959), *Portfolio Selection* (Cowles Foundation Monograph No. 16). John Wiley & Sons, Inc., New York.

Martin, D. W., and G. J. Tee (1961), "Iterative Methods for Linear Equations With Symmetric Positive Definite Matrix," *The Computer Journal*, IV, 242–54.

McCormick, G. P., and W. I. Zangwill (1967), "A Technique for Calculating Second-Order Optima," Research Analysis Corp., McLean, Virginia (mimeo.).

Michael, E. (1951), "Topologies on Space of Sets," *Transactions American Mathematics Society*, LXXI, 151–82.

Miller, C. (1963), "The Simplex Method for Local Separable Programming," Chap. 12 in *Recent Advances in Mathematical Programming*. eds. R. Graves and P. Wolfe, McGraw-Hill Book Company, New York.

Nelder, J. A., and R. Mead (1965), "A Simplex Method for Function Minimization," *The Computer Journal*, VIII, No. 3, 308–13.

Nemhauser, G. L. (1966), *Introduction to Dynamic Programming*. John Wiley & Sons, Inc., New York.

Nikaido, H. (1954), "On von Neumann's Minimax Theorem," *Pacific Journal of Mathematics*, IV, 65–72.

Oliver, L. T., and D. J. Wilde (1964), "Symmetric Sequential Minimax Search for a Maximum," *Fibonacci Quarterly*, II, No. 3, 169–75.

Penrose, R. (1955), "A Generalized Inverse for Matrices," *Proceedings of the Cambridge Philosophical Society*, LI, 406–13.

Pietrzykowski, T. (1962), "Application of the Steepest Descent Method to Concave Programming," pp. 185–89 in *Proceedings of the IFIPS Congress*, North Holland Company, Amsterdam, Holland.

Pisano, Leonardo (Fibonacci) (1857), *Scritti*, I, 283–84.

Polak, E., and M. Deparis (1967), "An Algorithm for Minimum Energy Control," Memo #ERL-M225, University of California, Berkeley.

Pomentale, T. (1965), "A New Method for Solving Conditioned Maxima Problems," *Journal of Mathematical Analysis and Application*, X, 216–20.

Pontryagin, L. S., V. G. Boltyanskii, R. V. Gamkrelidze, and E. F. Mischenko (1962), *The Mathematical Theory of Optimal Processes*. Translated by K. K. Trinogoff. John Wiley & Sons, Inc., New York.

Powell, M. J. D. (1962), "An Iterative Method for Finding Stationary Values of a Function of Several Variables," *The Computer Journal*, V, No. 2, 147.

―――― (1964), "An Efficient Method for Finding the Minimum of a Function of Several Variables without Calculating Derivatives," *The Computer Journal*, VII, No. 2, 155–62.

Reiter, S. (1954), *Efficiency and Prices in the Theory of an International Economy*, Technical Report No. 13, Stanford University, Stanford, Calif.

Riley, V., and S. I. Gass (1958), *Linear Programming and Associated Techniques*. Johns Hopkins Press, Baltimore, Md.

Rissanen, J. (1966), "On Duality Without Convexity," RJ389, IBM San Jose Research Laboratory, San Jose, Calif.

Ritter, K. (1965), "Stationary Points of Quadratic Maximum Problems," *Z. Wahrscheinlichkeitstheorie Verw*, Geb 4, 149–58.

―――― (1966), "A Method for Solving Maximum Problems with a Nonconcave Quadratic Objective Function," *Z. Wahrscheinlichkeitstheorie Verw*, Geb 4, pp. 340–51.

Rockafeller, R. T. (1960), "Duality Theorems for Convex Functions," *Bulletin of the American Mathematics Society*, LXX, 189–92.

―――― (in Press), *Convex Analysis*. Princeton University Press, Princeton, N.J.

Rosen, J. B. (1960), "The Gradient Projection Method for Nonlinear Programming, Part I, Linear Constraints," *Journal of the Society for Industial and Applied Mathematics*, VIII, No. 1, 181–217.

―――― (1961), "The Gradient Projection Method for Nonlinear Programming, Part II, Nonlinear Constraints," *Journal of the Society for Industrial and Applied Mathematics*, IX, No. 4, 514–32.

———— (1963), "Convex Partition Programming," pp. 159–76 in *Recent Advances in Mathematical Programming*. eds. R. L. Graves and P. Wolfe, McGraw-Hill Book Company, New York.

———— (1965), "Existence and Uniqueness of Equilibrium Solutions for Concave n-Person Games," *Econometrica*, XXXIII, No. 3, 520–34.

Rosen, J. B., and J. C. Ornea (1963), "Solution of Nonlinear Programming Problems by Partitioning," *Management Science*, X, No. 1, 160–73.

Rosenbrock, H. H. (1960), "An Automatic Method for Finding the Greatest or Least Value of a Function," *The Computer Journal*, III, No. 2, 175.

Royden, H. L. (1963), *Real Analysis*. The MacMillian Company, Publishers, New York.

Rutenberg, D. (1967), "Large Special Matrix Structures and the Convex Simplex Method," Working Paper No. 216, Center for Research in Management Science, University of California, Berkeley.

Saaty, T. L., and J. Bram (1964), *Nonlinear Mathematics*. pp. 168–70, McGraw-Hill Book Company, New York.

Samuelson, P. (1938), "A Note on the Pure Theory of Consumers Behavior," *Econometrica*, XVII, 61–71 and 353–54.

———— (1947), *Foundations of Economic Analysis*. Harvard University Press, Cambridge, Mass.

Sanders, J. L. (1965), "A Nonlinear Decomposition Principle," *Operations Research*, XIII, No. 2, 266–71.

Schecter, S. (1962), "Iteration Methods for Nonlinear Programming," *Transactions American Mathematical Society*, CIV, 179–89.

Shah, B. V., R. J. Buehler, and O. Kempthorne (1961), "The Method of Parallel Tangents (Partan) for Finding an Optimum," Office of Naval Research Report, NR-042-207, No. 2.

———— (1964), "Some Algorithms for Minimizing a Function of Several Variables," *Journal of the Society for Industrial and Applied Mathematics*, XII, No. 1, 74–92.

SIAM Journal on Control (1966), See particularly IV, No. 1.

Simonnard, M. (1966), *Linear Programming*. Translated by W.S. Jewell. Prentice-Hall, Inc., Englewood Cliffs, N.J.

Simmons, G. F. (1963), *Topology and Modern Analysis*. McGraw-Hill Book Company, New York.

Slater, M. (1950), "Lagrange Multipliers Revisited: A Contribution to Nonlinear Programming," Cowles Commission Discussion Paper Math. 403.

Spang, H. A. (1962), "A Review of Minimization Techniques for Nonlinear Functions," *Journal of the Society for Industrial and Applied Mathematics*, IV, No. 4, 343–65.

Stiefel, E. (1955), "Relaxationsmethoden bester Strategie zur Losung linearer Gleichungssysteme," *Commentarii Math.*, Helvetici 29, pp. 157–79.

Stoer, J. (1963), "Duality in Nonlinear Programming and the Minimax Theorem," *Numerische Mathematik*, V, 371–79.

Stone, R. (1964), "Linear Expenditure Systems and Demand Analysis: An Application to the Pattern of British Demand," *The Economic Journal*, LXIV, 511–27.

Stong, R. E. (1965), "A Note on the Sequential Unconstrained Minimization Technique for Non-Linear Programming," *Management Science*, 12, I, No. 12, 142–44.

Theil, H., and C. van de Panne (1960), "Quadratic Programming as an Extension of Classical Quadratic Maximization," *Management Science*, VII, No. 1, 1–20.

Tinter, G. (1955), "Stochastic Linear Programming with Applications to Agricultural Economics," in *Second Symposium in Linear Programming*, pp. 197–228.

Topkis, D. M., and A. F. Veinott, Jr. (1967), "On the Convergence of Some Feasible Direction Algorithms for Nonlinear Programming," *SIAM Journal on Control*, V, No. 2, 268–79.

Vajda, S. (1961), *Mathematical Programming*. Addison-Wesley Publishing Co., Inc., Reading, Mass.

van de Panne, C., and A. Whinston (1964), "The Simplex and Dual Method for Quadratic Programming," *Operational Research Quarterly*, XV, 355–88.

Van Slyke, R. M., and R. J. B. Wets (1966), "L-Shaped Linear Programs with Applications to Optimal Control and Stochastic Programming," ORC-66-17, Operations Research Center, University of California, Berkeley.

Veinott, Jr., A. F. (1967), "The Supporting Hyperplane Method for Unimodal Programming," *Operations Research*, XV, No. 1, 147–52.

Wagner, H. (1969), *Principles of Operations Research*. Prentice-Hall, Inc., Englewood Cliffs, N.J.

Warga, J. (1963a) "A Convergent Procedure for Convex Programming," *Journal of the Society for Industrial and Applied Mathematics*, XI, No. 1, 579–87.

——— (1963b), "Minimizing Certain Convex Functions," *Journal of the Society for Industrial and Applied Mathematics*, XI, No. 1, 588–93.

Wegner, P. (1960), "A Nonlinear Extension of the Simplex Method," *Management Science*, VII, No. 1, 43–50.

Wets, R. (1966), "Programming under Uncertainty: The Equivalent Convex Program," *SIAM Journal on Applied Math*, XIV, 89–105.

White, W. B., S. M. Johnson, and G. B. Dantzig (1958), "Chemical Equilibrium in Complex Mixtures," *Journal of Chemical Physics*, XXVIII, 751–55.

Wilde, D. J. (1964), *Optimum Seeking Methods*. Prentice-Hall, Inc., Englewood Cliffs, N.J.

Wilde, D. J., and C. S. Beightler (1967), *Foundations of Optimization*. Prentice-Hall, Inc., Englewood Cliffs, N.J.

Wilson, R. B. (1963), "A Simplicial Algorithm for Concave Programming." Unpublished Ph. D. dissertation, Harvard University Graduate School of Business Administration, Boston.

Wolfe, P. (1959), "The Simplex Method for Quadratic Programming," *Econometrica*, XXVII, No. 3, 382–98.

——— (1961a), "Accelerating the Cutting Plane Method for Non-linear Programming, "*Journal of The Society for Industrial and Applied Mathematics*, IX, 481–88.

——— (1961b), "A Duality Theorem for Nonlinear Programming," *Quarterly of Applied Mathematics*, XIX, 239–44.

——— (1962), "An Extended Simplex Method," *Notices of the American Mathematics Society*, IX, No. 4, 308 (Abstract).

——— (1966), "On the Convergence of Gradient Methods Under Constraints," *IBM Research Report RZ-204*, IBM Zurich Research Laboratories, Ruschliken, Zurich, Switzerland.

Zangwill, W. I. (1966a), "Minimum Concave Cost Flows, in Certain Networks," *Management Science—A*, XIV, No. 7, 429–50.

——— (1966b), "The Shortest Route Problem Under Either Concave or Convex Costs," Presented at 12th Annual Operations Research Society of American Meeting, Santa Monica, Calif.

——— (1966c), "A Deterministic Multi-Product, Multi-Facility Production and Inventory System," *Operations Research*, XIV, No. 3, 486–508.

——— (1966d), "A Backlogging Model and a Multi-Echelon Model of a Dynamic Economic Lot Size System—A Network Approach," Working Paper # 177, Center for Research in Management Science, University of California, Berkeley, Calif.

——— (1966e), "A Deterministic Multi-Period Production Scheduling Model with Backlogging," *Management Science—A*, XIII, No. 1, 105–19.

——— (1966f), "Production Smoothing of Economic Lot Sizes with Non-Decreasing Requirements," *Management Science—A*, XIII, No. 3, 191–209.

——— (1966g), "Convergence Condition for Nonlinear Programming Algorithms," Working Paper No. 197, Center for Research in Management Science, University of California, Berkeley, Calif.

——— (1967a), "Non-linear Programming via Penalty Functions," *Management Science—A*, XIII, No. 5, 344–58.

——— (1967b), "Extensions of Concavity, Minimax, Duality, and Optimality," Working Paper No. 219, Center for Research in Management Science, University of California, Berkeley.

——— (1967c), "The Piecewise Concave Function," *Management Science—A*, XIII, No. 11, 900–12.

——— (1967d), "Applications of the Convergence Conditions," 6th Symposium on Mathematical Programming, Aug. 14–18, 1967, Princeton University, Princeton, N.J.

——— (1967e), "An Algorithm for the Chebyshev Problem—with an Equivalence to Non-linear Programming," *Management Science—A*, XIV, No. 1, 58–78.

——— (1967f), "Minimizing a Function Without Calculating Derivatives," *The Computer Journal*, X, No. 3, 293–96.

——— (1967g), "The Convex Simplex Method," *Management Science—A*, XIV, No. 3, 221–38.

———— (1967h), "A Decomposable Nonlinear Programming Approach," *Operations Research*, XV, No. 6, 1068–87.

Zoutendijk, G. (1960), *Methods of Feasible Directions*. Amsterdam, Elsevier Publishing Co., Amsterdam, Holland.

———— (1966), "Nonlinear Programming: A Numerical Survey," *SIAM Journal on Control*, IV, No. 1, 194–210.

Zuhovickii, S. I., R.A. Poljak, and M.E. Primak (1963), "An Algorithm for the Solution of a Problem of Convex Cebysev Approximation," *Soviet Mathematics*, IV, No. 4, 901–4.

———— (1963), "An Algorithm for the Solution of the Convex Programming Problem," *Soviet Mathematics*, IV, No. 6, 1754–57.

Index

Z